The Unsung Heroes of World War I

WAR-HORSES

By G. M. JEUDWINE

WE combed you out from happy silences
    On thymey downs;
From stream-veined meadowlands alight with crowns
Of buttercups, where, for you, shapely trees
    Made spacious canopies.

Now (day and night) unsheltered, in the mud
    You droop and ache;
While ruthless hands, for human purpose' sake,
Fashion the complex tools which spill your blood
    And ours in rising flood.

No deputation (yet) your wage controls.
    Ungauged, unpaid
Your overtime. The war blast leaves no blade
Of green for you—poor ghosts of happy foals!—
    Munching your minished doles
In ravages by human frenzy made.

The Unsung Heroes of World War I
By: Gloria Austin
President of:
Equine Heritage Institute, Inc. (EHI)
and Mary Chris Foxworthy
with special thanks to Stephanie Sutch

First Publish Date 2018
Copyright © 2018 by Equine Heritage Institute, Inc.
ISBN: ISBN: 978-1-7320805-1-5

All rights reserved. No part of this publication may be reproduced, distributed, or transmitted in any form or by any means, including photocopying, recording, or other electronic or mechanical methods, without the prior written permission of the publisher, except in the case of brief quotations embodied in critical reviews and certain other noncommercial uses permitted by copyright law. For permission requests, write to the publisher, addressed "Attention: Permissions Coordinator," at the address below.

Gloria Austin Carriage Collection, LLC; Equine Heritage Institute, Inc.
3024 Marion County Road Weirsdale, FL 32195 Office: (352) 753-2826 Fax: (352) 753-6186

Ordering Information:
Quantity sales: Special discounts are available on quantity purchases by corporations, associations, and others. For details, contact the publisher at the address above.
Printed in the United States of America First Edition ISBN

The Unsung Heroes of World War I

"We have had 6,000 years of history with the domesticated horse and only 100 years with the automobile."

Gloria Austin

Gloria Austin and Mary Chris Foxworthy

THE HORSE HAS IMPACTED ALL SUSTAINABLE CULTURES AND CIVILIZATIONS AROUND THE WORLD

TABLE OF CONTENTS

Forward 7

The World at War 9
    What Caused the War? 10
    Who Was Fighting and Why? 13
    Where was the fighting? 16

Modern Warfare 21

The Unsung Heroes 23
    Obtaining Horses and Mules For The War 24
    How Were Horses and Mules Used in the War? .. 32
    The Cavalries of the War 40
    Horses and Mules on the Fronts 46

Care of Horses and Mules 55

Real Stories of Horse and Mule Heroes 66

The Homefront Without Horses and Mules 94

Armistice and the Post-War World 97
    What Ended the war? 98
    Bringing the Boys and the Horses Home 100
    Dorothy Brooke 104
    Devastation 106
    The Changing World—Goodbye to the Horse .. 107
    Horses Are Still Our Heroes 108

Modern-Day Horse Soldiers 109
    Wounded Warriors 111

Facts and Figures 112

Sources 117

Forward

During her tenure as the leader of the Gloria Austin Carriage Museum, many questions came up about horses in war. As Gloria researched these questions, she found a story that needed to be told. Questions like "What did horses eat in World War One?" and "Why were horses so important in World War One?" were often left unanswered. Wars start by leaders and are fought by soldiers and horses, or that was the case before the Korean Conflict.

While the propaganda of the day was exposing various causes for World War I, history shows that without horses and donkeys, the First World War could never have been fought. Combatants needed horses for transporting everything from artillery to injured soldiers.

In order to give this story the significant it deserved, Gloria brought in a researcher with knowledge of history and horses; Mary Chris Foxworthy. Together they collaborated to create this book. A look at the Unsung Heroes of World War One.

Research into the events of the war brought to light the vital need for equine support during the war. Casualties would have been incalculably higher, weapons would never have arrived and the allies would never have gotten the upper hand without our equine contributors. Within this book you will find true heroes from the general's horse to the ammunition supply donkey, all acting above and beyond the call of duty to keep soldiers alive.

Gloria Austin

Gloria Austin is a world renowned horse historian and is particularly well known as a carriage driver and collector of fine antique carriages. She holds many championship titles in single, pair and four-in-hand driving. She has driven over 17 different breed and has driven horses in the US, Canada, Europe, and Australia. She holds the North American Four-In-Hand and Coaching Champion in Pleasure Driving in the USA. As an accomplished presenter, Gloria entertains audiences with stories of riding, carriage driving, horses in history, and her worldwide travels visiting horse venues around the world. The social history of the horse has long been Gloria's passion. As we are reminded in one of her informative lectures, "We have had over 6000 years of history with the domesticated horse and only 100 years with the automobile." She is actively publishing books, many of which focus on the 6000 years when man used the horses in warfare, transportation, industry, agriculture, and commerce.

Mary Chris Foxworthy

Mary Chris' grandfather owned one of the last creameries in the United Sates that still used horse drawn milk wagons; thus began a life-long love affair with horses. After graduating from college with a degree in Food Science and Communications, Mary Chris bought her first horse with her first pay check. After graduation she worked in Marketing and Finance in the corporate world. During that time, she volunteered her Marketing and Fundraising skills for various non-profits and finally in 2006 left the corporate world to work in the area of Advancement until her retirement in 2016. She has served on the board of various equine associations and held a judge's card in Carriage Driving. She has published and presented numerous equine educational programs, written for several equine publications and won an award from American Horse Publications for one of her articles. Mary Chris is an active exhibitor in Carriage Driving and Dressage. Along with her husband, she enjoys spending time with their Morgan horses, bouncing Bearded Collie and two adult children.

INTRODUCTION - Warfare with Horses Before World War I

The first use of horses in warfare occurred over 5,000 years ago. The earliest evidence of horses ridden in warfare dates from Eurasia between 4000 and 3000 BC. A Sumerian illustration of warfare from 2500 BC depicts some type of equine pulling wagons. By 1600 BC, improved harness and chariot designs made chariot warfare common throughout the Ancient Near East. The earliest written training manual for war horses was a guide for training chariot horses written about 1350 BC. As formal cavalry tactics replaced the chariot, so did new training methods, and by 360 BC, the Greek cavalry officer Xenophon had written an extensive treatise on horsemanship. The effectiveness of horses in battle was also revolutionized by improvements in technology, including the invention of the saddle, the stirrup, and later, the horse collar.

Horses were well suited to the warfare tactics of the nomadic cultures of Central Asia. Several East Asian cultures extensively used cavalry and chariots. Muslim warriors relied upon light cavalry in their campaigns throughout North Africa, Asia, and Europe beginning in the 7th and 8th centuries AD. Europeans used several types of war horses in the Middle Ages, and the best-known heavy cavalry warrior of the period was the armored knight.

With the decline of the knight and rise of gunpowder in warfare, light cavalry again rose to prominence, used in both European warfares and in the conquest of the Americas. Battle cavalry developed to take on a multitude of roles in the late 18th century and early 19th century and was often crucial for victory in the Napoleonic wars. In the Americas, the use of horses and development of mounted warfare tactics were learned by several tribes of indigenous people, and in turn, highly mobile horse regiments were critical in the American Civil War.

(cited from http://enacademic.com/dic.nsf/enwiki/4054154)

THE USE OF HORSES IN WARFARE WAS ABOUT TO CHANGE......

The World at War

What Caused the War?

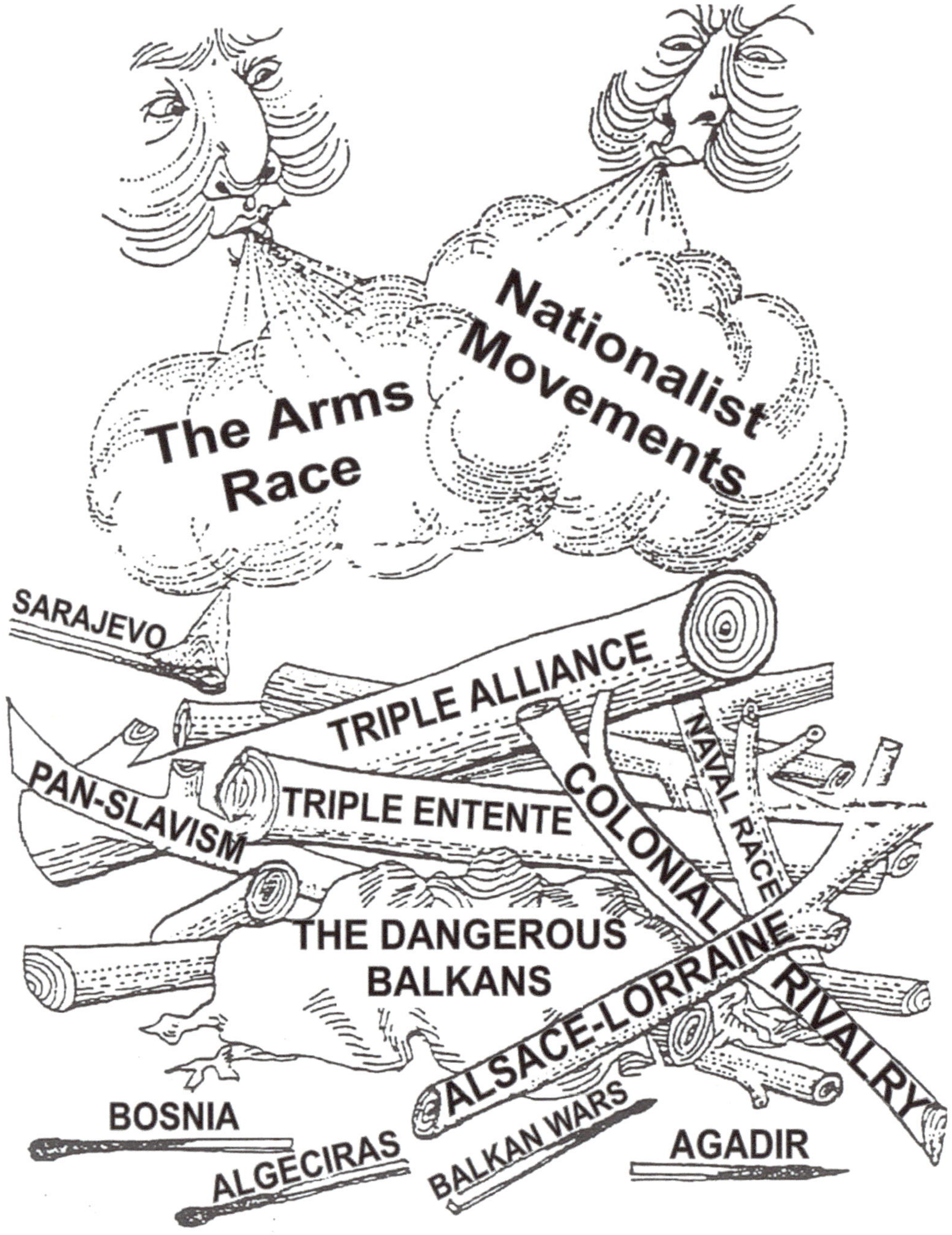

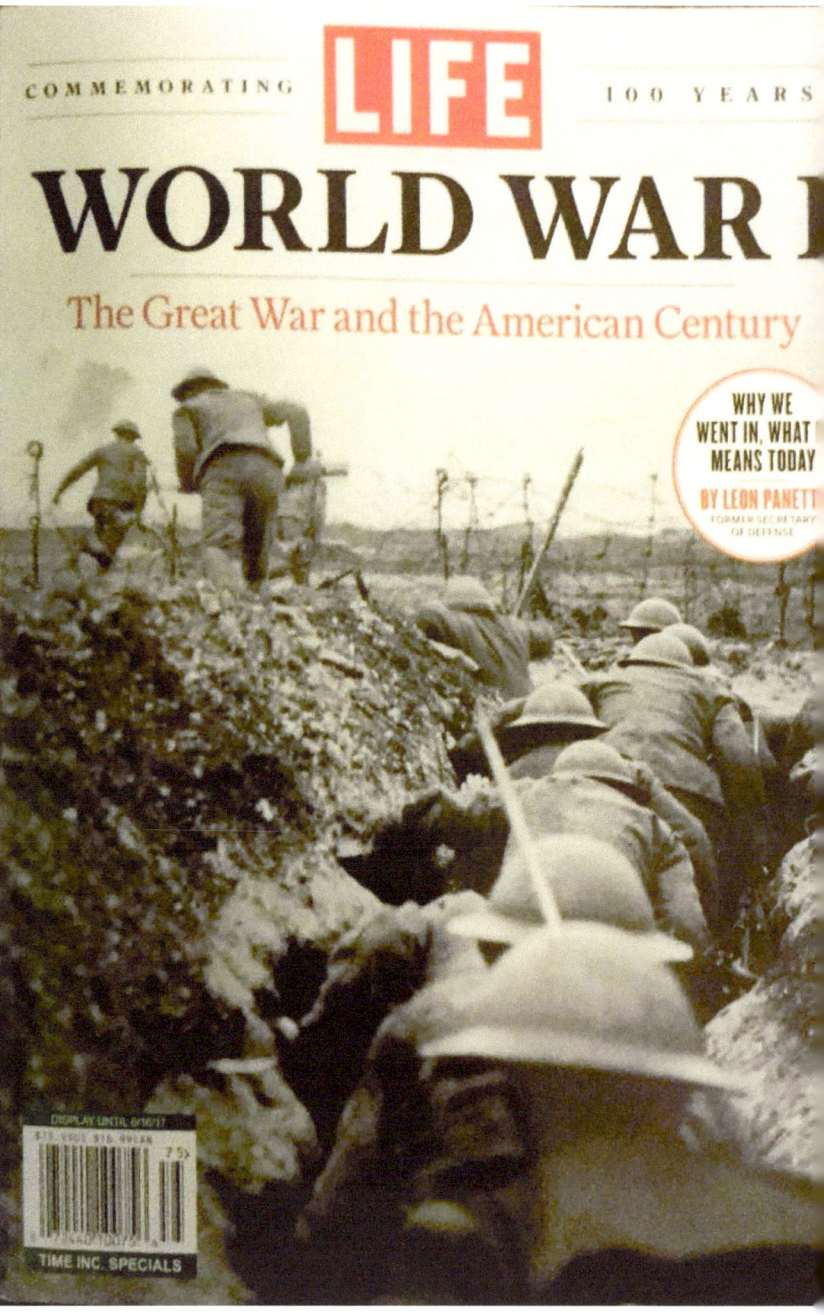

The Unsung Heroes of World War I

World War I lasted from July 28,1914 to November 11, 1918. Often called "The Great War" and referred to as "the war to end all wars"; it was a complex war with many combatants. More than 9 million soldiers and civilians died, and 20 million were injured. The M-A-I-N acronym is often used to analyze the war–militarism, alliances, imperialism, and nationalism.

MILITARISM: This was an era of military competition which created a culture of paranoia that heightened the search for alliances.

ALLIANCES: A web of alliances created two camps: The Triple Alliance of 1882 linked Germany, Austria-Hungary and Italy. The Triple Entente of 1907 linked France, Britain and Russia. A point of conflict between Austria Hungary and Russia was over their incompatible Balkan interests, and France suspected Germany rooted in their defeat in the 1870 war.

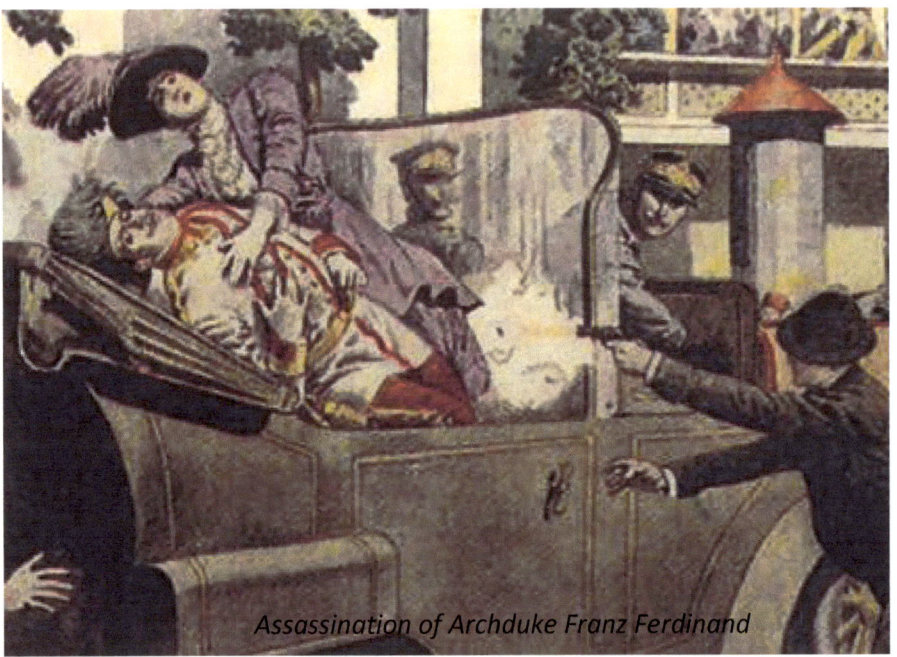
Assassination of Archduke Franz Ferdinand

IMPERIALISM: Imperial Competition also pushed the countries towards adopting alliances. Colonies were units of exchange that could be bargained without significantly affecting the metro-pole.

NATIONALISM: Nationalism created new areas of interest over which nations could compete. The Habsburg empire was an agglomeration of 11 different nationalities, with large Slavic populations in Galicia and the Balkans whose nationalist aspirations ran counter to imperial cohesion. Serbian nationalism created the trigger cause of the conflict–the assassination of the heir to the Austro-Hungarian throne–Archduke Franz Ferdinand.

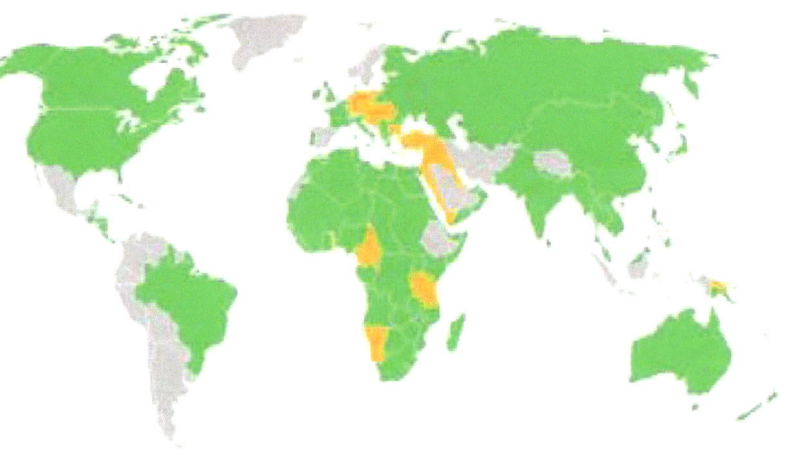
Allies   Central Powers   Neutral

The Schlieffen plan could be blamed for bringing Britain into the war. Blame has been directed at every single combatant at one point or another. Every point has some merit. What proved most devastating was the combination of an alliance network with the widespread, misguided belief that war is good for nations, and that the best way to fight a modern war was to attack

The Unsung Heroes of World War I

The Schlieffen plan could be blamed for bringing Britain into the war. The Schlieffen Plan was Orchestrated by Count Alfred von Schlieffen. It called to wage a two-front war due to the Franco Russian Alliance of 1894. Germany would take advantage of its central position and outstanding striking power of its army to take out France first, then turn its attention to Russia.

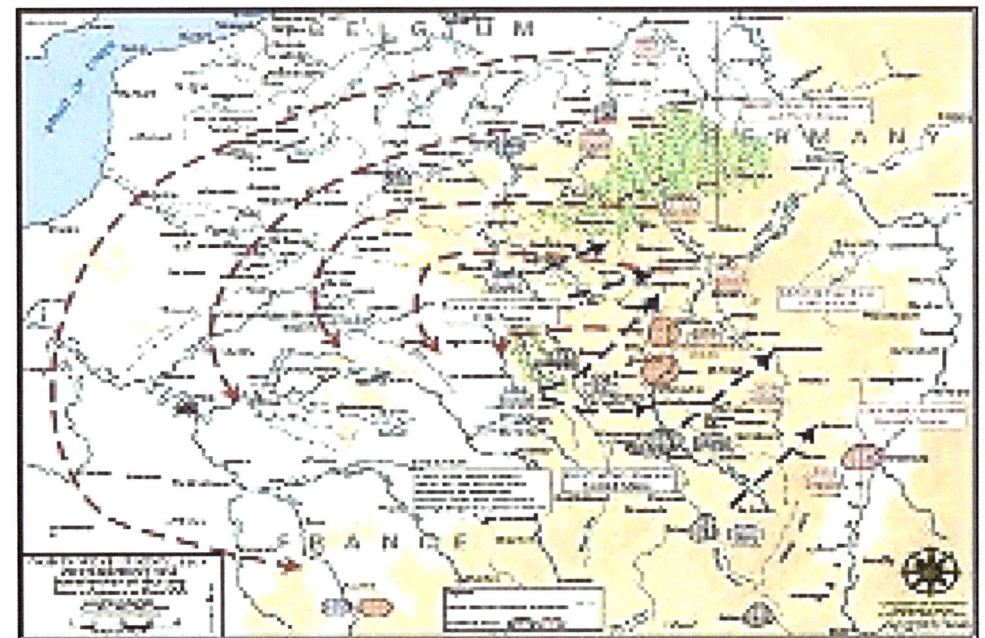

The order of events that started World War I are:

- Austria Hungary attacked Serbia
- Russia defended Serbia
- Germany attacked Russia to defend Austria
- France honored their alliance with Russia and declared war on Germany
- Germany invaded Belgium, violating Belgian neutrality, which brought in the UK and all the commonwealth.
- 90% of Africa was under the French, British, Germans and Belgians, so they fought too.
- The Germans sank the Lusitania ship, so the USA joined in.
- Japan and China were British allies and attacked German colonies in the Pacific.
- The Ottomans wanted to take advantage of Russian weakness to attack the Caucasus, so they allied Austria Hungary.
- The Balkans were in the middle of the attention, so Bulgaria also joined in against Serbia for revenge on the Balkan wars.
- Italy saw an opportunity to reconquer Italian lands under Austrian control, so they joined the war against the Germans.
- The rest of those involved were either under British or French influence, so they joined the war for them.

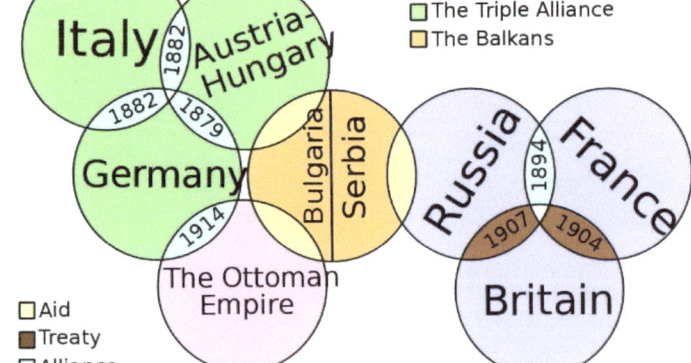

(Cited from: https://www.quora.com/Why-were-other-countries-involved-in-WW1)

- **Entente Powers (Allies)**
  - Russian Empire
  - France
  - British Empire
  - Italy
  - United States
  - Greece
  - Belgium
  - Japan
  - Romania
  - Montenegro
  - Serbia
  - Portugal

- **Central Powers**
  - German Empire
  - Austria-Hungry
  - Ottoman Empire
  - Bulgaria

When Germany invaded Belgium, both sides felt that the conflict would be short and over by Christmas since urban societies relied on the food and raw materials of a world economy. They had no idea what was about to come.

German field artillery advancing towards France

"The boys will be home by Christmas."

The Unsung Heroes of World War I

Where was the fighting?

The Western Front

The German army crossed the Belgian border in August 1914. Britain and France declared war on Germany on August 4th. The Germans pushed through Belgium and entered France. By the end of August, the French Army had suffered 75,000 dead of which 27,000 were killed on August 22nd alone. Total French casualties for the first month of the war were 260,000 of which 140,000 were sustained during the climactic final four days of the Battle of the Frontiers. The French still used outdated military tactics of marching in a tight formation towards enemy lines.

The British and French armies marched to stop the German advance. The Battle of Marne September 4 -10, 1914 stopped the Germans from marching on Paris. The war became a stalemate when the Allies won the Battle of the Marne. It was the second major clash on the Western Front and one of the most important single events of the war. The German retreat left the Schlieffen Plan in ruins and German hopes of a quick victory in the west. Its army was left to fight a long war on two fronts. Over two million men fought in the First Battle of the Marne, of whom over 500,000 were killed or wounded. Both sides dug in, and trench warfare began. Throughout the entire war, neither side gained more than a few miles of ground along what became known as the Western Front.

To protect themselves from machine gun fire and mustard gas, both sides dug trenches and lined them with barbed wire and mines. The trenches extended from the Belgium ports to the Swiss frontier. The horrors of trench warfare included mud, rats and rotting bodies. Throughout the entire war, neither side gained more than a few miles.

The Battle of the Somme, also known as the Somme Offensive, was one of the largest battles of the First World War. Fought between July 1 and November 1, 1916, near the Somme River in France, it was also one of the bloodiest military battles in history; 1,100,000 men were killed or wounded.

Trenches on the Western Front

The Unsung Heroes of World War I

The Eastern Front

The line of fighting on the Eastern side of Europe between Russia and Germany and Austria-Hungary is known as the Eastern Front. Fighting began on the Eastern front when Russia invaded East Prussia on August 17, 1914. Germany immediately launched a counter-offensive and pushed Russia back. This pattern of attack and counter-attack continued for the first two years of the war and meant that the Eastern Front changed position as the land was captured and lost by both sides.

2.5 million Russians were killed, wounded or taken prisoner. By 1917, the Russian people were fed up with the huge number of Russian losses. The government and monarchy were overthrown, and the new Bolshevik government signed the treaty of Brest Litovsk which took the Russians out of the war.

Trench warfare did not fully develop on the Eastern Front.

Photos from the Eastern Front

The Italian Front

Before the outbreak of war in August 1914, Italy had sided with Germany and Austria-Hungary. However, tempted by offers of more land once the war was won, Italy entered the war in April 1915 on the side of the allies. The Italian front is the name given to the fighting that took place along the border between Italy and Austria. The Italians only advanced a short way into Austria. Between 1915 and 1917 there were twelve battles fought along the river Isonzo just inside the Austrian border. After being defeated at the battle of Caporetto, the Italians were pushed back.

Official war photographers Ernest Brooks and William Joseph Brunell found beauty and brutality on the Italian front.

Gallipoli

The Gallipoli peninsula is located in the south of Turkey. In 1915, the allied commanders tried to attack Germany by attacking her ally, Turkey. Allied soldiers, mainly from Australia and New Zealand, were sent to the peninsula while British ships tried to force a way through the Dardanelles. The entire mission was a failure. The allies lost over 50,000 men but gained hardly any land.

(Cited from https://schoolworkhelper.net/where-was-world-one-fought-theatres-of-wwi/)

Horse from the Gallipoli campaign

The Unsung Heroes of World War I

The Middle East

The Ottoman Empire, comprising present-day Turkey, Syria, Palestine, Iraq, Jordan, and parts of Saudi Arabia and Armenia, was a major power in the Middle East at the beginning of the war. The Ottoman Empire joined the Central Powers late in 1914 after the secret Ottoman-German Alliance was signed. In the Sinai and Palestine, hostilities between the Allied Powers (primarily Britain and Russia) and the Central Powers (primarily the Ottoman Empire and Germany) began in 1915 when the Ottomans launched an unsuccessful attack across the Sinai to try to capture the Suez Canal, threatening Russian and British territories and communication. After another unsuccessful Ottoman attack in 1916, the British went on the offensive, attacking into Palestine. Late in 1917, the British captured Gaza and Jerusalem. Hostilities officially ended on October 30, 1918, with the signing of the Armistice of Mudros and, shortly after that, the Ottoman Empire was dissolved, and the Turkish War of Independence began. During the years of

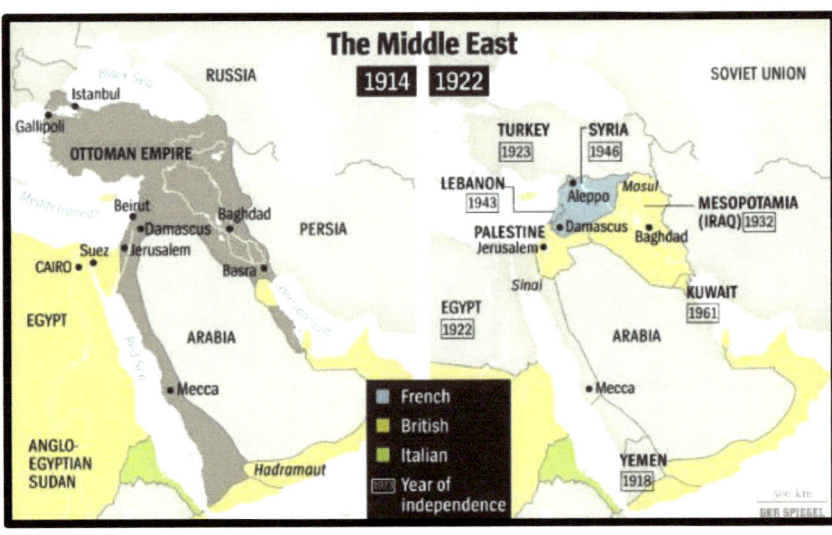

the Sinai and Palestine campaign, while the British worked overtly to occupy parts of the Ottoman Empire, they also worked covertly to incite a revolt among the Arabs living in present-day Saudi Arabia. The revolt against the Ottoman forces began in 1916 and was, in large part, planned and directed by Lieutenant-Colonel Thomas Edward Lawrence of the British army, better known as Lawrence of Arabia.

A Yeomanry patrol and their horses during a pause in the desert. The horses are equipped with nosebags and fringed browbands to protect their eyes from the sun, sand and insects.

(Cited from
http://www.militarian.com/threads/wwi-sinai-and-palestine-campaign.7252/)

Horses and camels were crucial to the advance. There were times they were ridden for more than 30 hours without water.

Modern Warfare

The Unsung Heroes of World War I

When ancient tactics like columns of armies marching towards each other and cavalry charges became overwhelmed by new weapons like machine guns, tanks, gas and air attacks, warfare changed forever. World War I was the first war in the industrialized age. New weapons and vehicles that have never been used in war were introduced.

The invention of the airplane in 1901 paved the way for the plane to be used as a gunship, a bomber and for reconnaissance. 124 different types of planes were used in World War I.

The tank was invented in World War I. The first tanks were awkward and were called "land ships." These first tanks had side-mounted cannons and turrets.

World War I was the first war in which poisonous gas was used. Chlorine gas was the first gas used then phosgene gas then mustard gas. The gas damaged nerves, caused blindness and more. The gas also stayed in the earth and poisoned the trenches. After the war, poisonous gas was outlawed internationally.

In 1915 the Germans began to use flamethrowers. They were cumbersome so not widely used.

The machine gun was introduced to the British army in 1884 by Hiram Maxim. The British officers had no use for it and felt it was an improper form of warfare. The Germans liked the idea and created their own version of the gun. At the start of the war, the Germans had 100,000 "maschinegewehrs" while the British only had a few hundred.

War had indeed changed. Men and horses were mowed down in droves every time they charged against positions fortified with barbed wire, trenches, automatic weapons, and tanks. As cavalry historian, Dr. Alexander Bielakowski stated, "You're going against machine guns with a long stick."

The Unsung Heroes of World War I

Obtaining Horses and Mules For The War

When the war began, the British army had a mere 25,000 horses. That may seem like a lot of horses but, considering that 8 million horses and countless mules and donkeys died in the war, it then seems hard to imagine how the various nations were able to obtain so many horses and mules. Horses were shipped from Spain, Portugal, New Zealand, South Africa, India, Canada, and America.

At one-point America was sending 1,000 horses a day. Beginning in late November 1914, the port of Newport News, Virginia became the biggest shipper of American war horses and mules to the British army in Europe in a crucial effort that helped the Allies win the war. Killing American horses and mules became a strategic priority for the Germans. A German sabotage campaign was the first attempt at using germs in warfare. Clandestine attempts by German agents to infect and kill the horses were not even discovered until after the war.

Horses ready for shipment at Newport News.

At the outbreak of the war, Americans owned between 24 and 25 million horses and 4½ million mules. There were 2 million horses in the United Kingdom, 3.2 million in France, 1 million in Italy, 4.5 million in Germany and 1.8 million in Austria. More than half the horses on earth lived in the United States and Russia. (from US Department of Agriculture Yearbook 1920 pp. 701-717, table 229) The U. S continued to have a robust horse and mule population throughout the war years due to the Federal Government encouraging peak production of foodstuffs and land grew supplies with the "*Food Will Win the War*" slogan.

Horses arriving in Europe by ship.

Cavalry remount post in America 1917.

The Unsung Heroes of World War I

Beginning in September 1914, U.S. horse exports rose rapidly, going from 804 in August to 7,146 in September, 12,091 in October, 28,071 in November and 30,687 in December. They reached a peak of 47,380 in July 1915 and did not fall below 18,000 a month until 1917. *(from Bureau of Agricultural Economics "Horses, Mules, and Motor Vehicles," January 1925 pp. 21-22 Table 28.)*

In 1915 alone, the U.S. exported 355,000 horses and mules to Europe Between 1916 and 1919 the U.S. exported 719,000 horses and 270,000 mules to the battlefields, supply lines, and farms of the allies.

Mule remounts at Camp Jackson, South Carolina

Captain Sidney Galtery, a Remount Officer, had the job of supervising the handling of horses in France from their arrival on ships through the casting of horses. His book. "The Horse and the War," is a detailed accounting of the War Horse in World War I. He commented on the American horses that were shipped to Europe that "… of all the breeds and cross-breeds of horses in the world the one from the United States and Canada has proved paramount and incomparably the best…. Hardiness, placidity of temper, strength, and power, virility of constitution, with what is called 'good heart,' versatility and extraordinary activity for his size and weight — these are characteristics that have impressed themselves for all time on all who have had to do with him… the light draught of American origin has come to stay in this country. After all, they are a distinct type. Some may be better than others, and some may be heavier in physique than the vast majority, but these latter are as if they had all come out of the same mold. By comparison, the British light draught is a nondescript, a misfit. He could be anything — a half-bred Shire or Clydesdale, a Welsh cob, a heavish Hackney, a Cleveland bay, or a heavy-weight 'hunter' without true hunter lines and action. All these odds and ends of horse-flesh we have seen pass through remount depots on route to the theatres of war. They were classed as light draught because they were neither heavy draught nor a riding horse. But the Yankee was essentially and absolutely a light draught horse, true to type, varying not at all in character and very little in the non-essential details. He is the real equine hero of the war, and by his triumphs, which must be as real in peacetime as in war, he simply must take his place, and an important one, too, in the horse population of these Islands."

Fort Nebraska remount station

Once America entered the war, the American Expeditionary Force (A.E.F.) also needed horses. On May 12 and June 17, 1917, Congress appropriated $53,000,000 ($1,126,359,655 in 1918 dollars) for the purchase of 250,000 horses and mules. *(from the report of the Quartermaster General to the Secretary of War, 1917).* By mid-1918 a million men were ready to go to Europe. Could the mules and horses they had trained go with them? General Pershing *(left)* was to serve as the commander of the American Expeditionary Force. He had experiences in the management of transporting mules and horses in the 1916-17 Mexican border campaign. The Mexican campaign, however, did not have to deal with overseas transport. The German submarine warfare that caused the U.S. entry into the war was making the transport of men, supplies, and horses very difficult. Shipbuilding increased in 1917, but twice as much tonnage was sunk as was built. Americans ended up going to Europe without the horses they needed. The French offered to supply Americans with 7,000 horses per month but soon found they could not do that and agreed to "loan" 4,000 draft horses for the American artillery expected to arrive. Interesting to think that perhaps these were horses that the U.S. had supplied in the first place. By December 1917, 12,414 animals had been provided by the French and 9,550 imported from the U.S. Other purchases and methods of requisition were attempted but the U. S still did not have the number of horses needed. Faced with these great shortages, the A.E.F. took steps to reduce the division allowance of animals by dismounting certain officers and men, substituting bicycles *(below)* for horses and motorizing certain units. The supply of horses to the A.E.F. remained inadequate through the end of the war.

General Pershing wrote: "In these last days of the fighting some troops, including the 81st, operated with a serious shortage of animals, which made it impossible to employ all their artillery in close support of the infantry, and often required men to drag their guns by hand."

The Unsung Heroes of World War I

THE REMOUNT TRAIN

By W. H. OGILVIE

EVERY head across the bar,
   Every blaze and snip and star,
Every nervous, twitching ear,
Every soft eye filled with fear,
Seeks a friend, and seems to say:
" Whither now, and where away ? "
Seeks a friend and seems to ask:
" Where the goal, and what the task ? "

Wave the green flag! Let them go!—
Only horses? Yes, I know;
But my heart goes down the line
With them, and their grief is mine!—
There goes honour, there goes faith,
Down the way of dule and death,
Hidden in the cloud that clings
To the battle-wrath of kings!

There goes timid child-like trust
To the burden and the dust!
High-born courage, princely grace
To the peril it must face!
There go stoutness, strength and speed
To be spent where none shall heed,
And great hearts to face their fate
In the clash of human hate!

Wave the flag, and let them go!—
Hats off to that wistful row
Of lean heads of brown and bay,
Black and chestnut, roan and grey!
Here's good luck in lands afar—
Snow-white streak, and blaze, and star!
May you find in those far lands
Kindly hearts and horsemen's hands!

The Field Artillery officers agreed that "our friend from America," the light draught, half-bred Percheron withstood the rigors and exhausting exposure of active service while the heavy horse of Britain succumbed. "The virtues of the type — great endurance, fine physique, soundness, activity, willingness to work, and almost unfailing good temper" made them the most desirable of all war horses. The extraordinary thing was that America was able to consistently supply this type of quality horses for 5 years. In purchasing such large numbers of animals, it was imperative to buy only from well-known and reliable horse dealers. Such dealers had their show-yards in large towns where the livestock business is a big concern. The chief centers used were Chicago, St. Paul in Minnesota, Sioux City and Des Moines in Iowa, St. Louis, and Kansas City.

Guyton & Harrington were mule dealers with facilities in Lathrop, Missouri, Kansas City, East St. Louis and Port Chalmette, Louisiana. The volume of mules they supplied was so great that 500 buyers were needed in Lathrop alone just to keep the flow running smoothly. They sold over 180,000 mules to the British army between 1914-1918.

Horses needed to be shipped from the Midwest to the shipyards of the east. 70% of the horses shipped by railway developed shipping fever. These horses were held until well before embarking on a ship. The trip across the Atlantic was not an easy one. Horses were kept in tight quarters. Many horses did not survive the journey and ships transporting horses were often the targets of German U-boats.

Britain requisitioned horses from British civilians. Horses were pulled from the coal mines and city streets for service. The London bus company sold their remaining horses to the army. Lord Kitchener, upon receiving a letter from P. L. and Freda Hewlett *(right)* in August 1914, asking him to spare their pony, Betty, ordered that no horses under 15 hands should be confiscated.

"In England's recent preparations for war, no sight was more interesting to the visitor during the first week of hurried mobilization than the commandeering of horses in the city streets. An officer charged with this duty, bearing a certain number of orders for horses, rode through the streets, and when a horse of the proper size, build and the color was found a halt was ordered, the driver being asked to unhitch in the name of the Royal Service. After a hurried examination of the horse which was usually a satisfactory one, the Government's brand of a broad arrow was burned on the animal's hoof, and the owner's name and address were taken, with the assurance of a subsequent settlement, if the Government's veterinarian accepted the horse. Without further ado the horse was thus taken was then the property of the Government, and the army officer rode away, leaving the former owner standing on the curbstone." *(From the Breeders Gazette, October 22, 1914 p.692)*

165,000 horses were requisitioned in 1914 alone

The British Army Remount Service, to improve the supply of horses for potential military use, provided the services of high-quality stallions to British farmers for breeding their broodmares. The already rare Cleveland Bay was almost wiped out by the war; smaller members of the breed were used to carry British troopers while larger horses were used to pull artillery. New Zealand found that horses over 15.2 hands fared worse than those under that height. Well-built Thoroughbreds of 15 hands and under worked well, as did compact horses of other breeds that stood 14.2 to 14.3 hands. Larger crossbred horses were acceptable for regular work with plentiful rations but proved less able to withstand short rations and long journeys. Riflemen with tall horses suffered more from fatigue, due to the number of times they were required to mount and dismount the animals. Animals used for draught work, including pulling artillery, were also found to be more efficient when they were of medium size with good endurance than when they were tall, heavy and long-legged.

1915 Stony Stratford, England - market square full of requisitioned horses

The Unsung Heroes of World War I

Soon, a regular supply of 1,000 horses a week were being landed at Bristol's King Edward Dock and transported the short distance to the remount depot at Shirehampton. The depot soon held as many as 5,000 horses, and with a constant stream of horses arriving from the USA, an overflow facility had to be found. The depot at Ormskirk, near Liverpool, was also full and therefore unable to accept these overseas imports. The sea around the approach to Southampton was a haven for U-Boats, making it far too dangerous a proposition to contemplate traveling the extra distance around the coast to this facility. The only practical recourse available to the army, therefore, involved further requisitions, this time of suitable farmyards within a reasonable traveling distance of Shirehampton.

Eventually, 12 mule depots were established on private farms and land, which had been requisitioned by the army, and these were used to provide rest, recuperation and training for the animals before shipment to Swaythling. One such center was at Bratton Court, Somerset. The compensation for the requisition of the land was 30 shillings a week (around £140/$195 at today's values). That doesn't sound a great deal but add to this that an additional two shillings and sixpence were given for every mule that arrived at Bratton Court. This made the final figure around £5,500/$7,650 per week at today's values, a considerable sum. Compensation was certainly one area where the army operated a very fair and efficient system with a figure of over 36 million pounds being spent on the horse-purchasing program during the war.

One center at Russley Park, Wiltshire, was run entirely by women. *(right)* Originally designated as a stud for "top brass" chargers, it never fully achieved this role and became predominantly a training and rehabilitation center for officers' cavalry horses. The women who ran the center were all upper-class young girls who caused shock and outrage amongst the local population by not only wearing trousers and dressing as men but by sitting astride the horses when riding them out rather than using the traditional side-saddle. The queen had, as late as 1913, tried to ban women riding astride in Hyde Park and, at the horse show held at Olympia in 1914, the king refused to watch any woman that came into the arena riding astride. The care of horses was just one of many roles that women were able to excel at during WWI. *(cited from http://nerdalicious.com.au/history/war-horses-britains-equine-army-of-the-first-world-war/)*

Germany had a more organized system before the war and had sponsored horse-breeding programs in preparation. These efforts caused Germany to increase the ratio of horses to men in the army, from 1 to 4 in 1870 to 1 to 3 in 1914. German horses were registered annually with the government in much the same way as army reservists.

There are few pictures of the Central Powers on the fronts. Most German newspapers and magazines at the time had illustrations rather than pictures. The reporting was more propaganda than actual reporting. It is suspected that the artists who provided the illustrations had never actually visited the front and were providing idealized rather than realistic pictures.

From 1914 to 1916, the Entente and Central Powers were more or less evenly matched when it came to the provision of war supplies such as weapons, ammunition, food rations, and horses. Unlike the Allies, however, the Central Powers were unable to import horses from overseas and so, during the war, they developed an acute horse shortage which contributed to their defeat by paralyzing artillery battalions and supply lines.

German cavalry officers were well prepared when the war started

Artist's impression of the German attack on Russian troops.

Kaiser Wilhelm posing with a horse—it is said he conducted all affairs from a saddle

The Unsung Heroes of World War I

How Were Horses and Mules Used in the War?

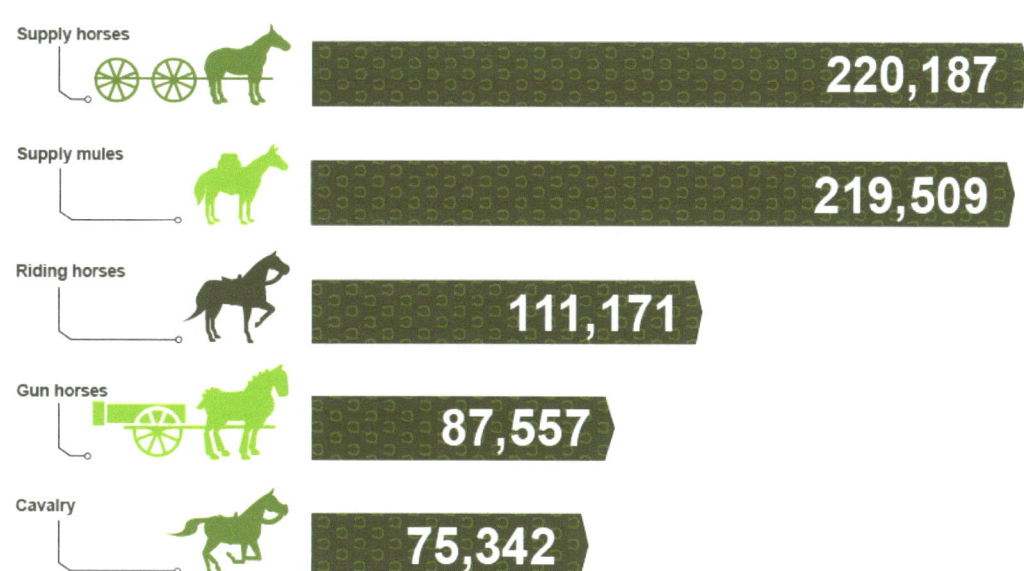

In 1914 the British army had only 80 motor vehicles, so the dependence on using horses and mules to transport goods and supplies was critical. The majority of horses were not used on the battlefield. In 1918 the British Army had almost 500,000 horses distributing 34,000 tons of meat and 45,000 tons of bread and 16,000 tons of forage for the horses each month.

(*cited from http://www.bbc.co.uk/guides/zp6bjxs*)

The thick mud rendered motorized vehicles useless in transporting supplies. Both sides relied on animal power for transporting everything from food to ammunition to ambulances. Many times, animals remained in draft 48–72 hours and were often caught in the open while delivering supplies.

British artillery delivering ammunition to the front

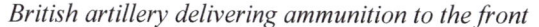

Royal artillery train moving guns

Field kitchen to feed the troops

The Unsung Heroes of World War I

> The power of an army as a striking weapon depends on its mobility. Mobility is largely dependent on the suitability and fitness of animals for army work.
>
> I hope that this account of our army horses and mules will bring home to the peoples of the British Empire and the United States the wisdom of breeding animals for the two military virtues of <u>hardiness</u> and <u>activity</u>, and I would add that the best animals for army purposes are also the most valuable for agriculture, commerce and sport.
>
> G.H.Q. France
> D. Haig, F.M.
> 19 Sep: 1918

Even during the war, few people realized the extent and importance of horses and mules to the war effort. Many believed that the world had entered an era of motor haulage and that horses were not necessary. Captain Sidney Galtrey explains below in his book, "The Horse and the War," how and where the horse was used. He wanted to inform people of the "vast and wonderful part played by our war horses." Referring to Captain Galtrey's book, Field Marshal Sir Douglas Haig *(commander of the British Expeditionary Force on the Western Front from late 1915 until the end of the war)* wrote the note on the left.

"What is the artillery that preponderates in modern warfare? The field gun, of course, which is the weapon of the Royal Field Artillery and Royal Horse Artillery. Each must have its own team of conditioned horses, and so when you count up the guns in a battery, the batteries in a brigade, the brigades in a division, the divisions in a Corps, and the Corps in our Armies on all the Fronts you arrive at a first calculation of the vital necessity of horses and mules in many tens of thousands, the wastage among which has to be watched with the greatest care in order that the establishments prescribed may be rigidly maintained. For easy mobility and flexibility in the rapid movement are vital and essential in the making of successful warfare. Then with the Artillery of every Division there must be a Divisional Ammunition Column, which means several hundred more animals, and again there is the Divisional Train Transport, chiefly horsed by weighty draught horses, while you must also bear in mind that every battalion of infantry has its own transport of at least half a hundred animals. Think also of the tremendous variety of other Units (especially those connected with Machine Guns and the Royal Engineers), which go to make an Army in being, each having horses or mules, or both, allotted to it. One has in mind Labor and Road Construction Companies, Railway Companies, Forestry Companies, units on Lines of Communication and the Medical Service. What of the cavalry? Jerusalem would never have been entered but for General Allenby's Cavalry; the crusade into the heart of Palestine was distinguished by the fine exploits of Yeomen of Warwickshire, Worcestershire, Buckinghamshire, and Berkshire; but for Indian cavalry Allenby's brilliant coup by which two Turkish armies were smashed would not have been possible; while the success and gallantry of the Dorsetshire Yeomanry at Matruh in the Senussi fill a sparkling page in near Eastern military operations. The advance to Baghdad and beyond along the shores of the Tigris was not made possible by guns and infantry alone. So, too, in France, when comparatively open warfare displaced the stalemate of trench warfare, we had cavalry coming into its own again."

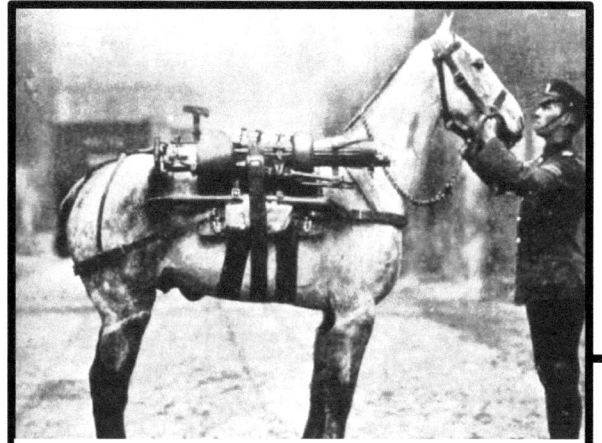

This horse is outfitted with a Vickers machine gun

Horses outfitted with gas masks.

Trapped while pulling in the deadly battlefield mud.

Mules transporting supplies

Pack horses transporting ammunition.

Russian Artillery

The Unsung Heroes of World War I

Reconnaissance

Horses were caught in the open while delivering supplies. Soldiers showed much empathy for their equine heroes and companions.

Supply train – US 129th Infantry. 33rd Division

To identify training levels; mule's tails were trimmed in a bell shape. A mule that was broke to pack, had a single bell trimmed into its tail. A drive broken mule had a second bell added below the first. Once broke to ride, a third tassel was trimmed below the second.

Mules and horses were vulnerable to getting stuck in the mud or falling into shell holes and drowning. Soldiers were quick to save them when possible.

Moving injured soldiers off the line.

More horses died from overexposure and disease than from enemy fire. Brigadier-General Frank Percy Crozier took part in the Battle of the Somme and said, "If the times are hard for human beings, on account of the mud and misery which they endure with astounding fortitude, the same may be said of the animals. My heart bleeds for the horses and mules."

Sir Douglas Haig's mount – note the horse's emaciated condition

Horses were picketed outside and susceptible to the elements.

Horses often sunk in the mud and were groomed and cared for by the soldiers when this happened.

The Unsung Heroes of World War I

The Cavalries of the War

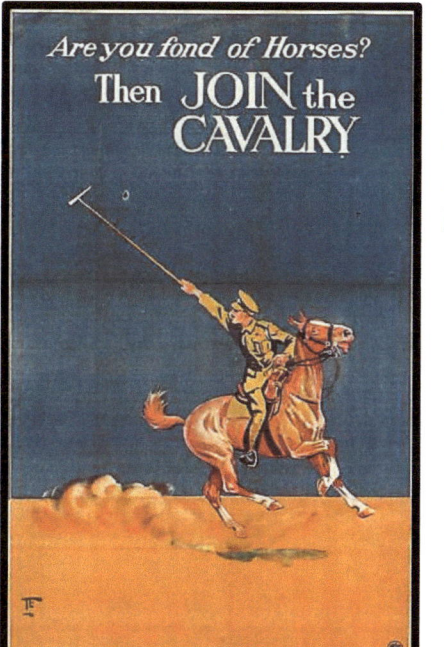

At the start of the war, the idea of war seemed romantic and noble. Joining a cavalry was the epitome of this idea. Once at the front, many of the cavalries were not needed. Horses stayed at depots while men were deployed to the infantry ranks. Many men were converted to communications or riflemen with the horses, or infantry without their horses and if horses were suitable, they were used for transportation.

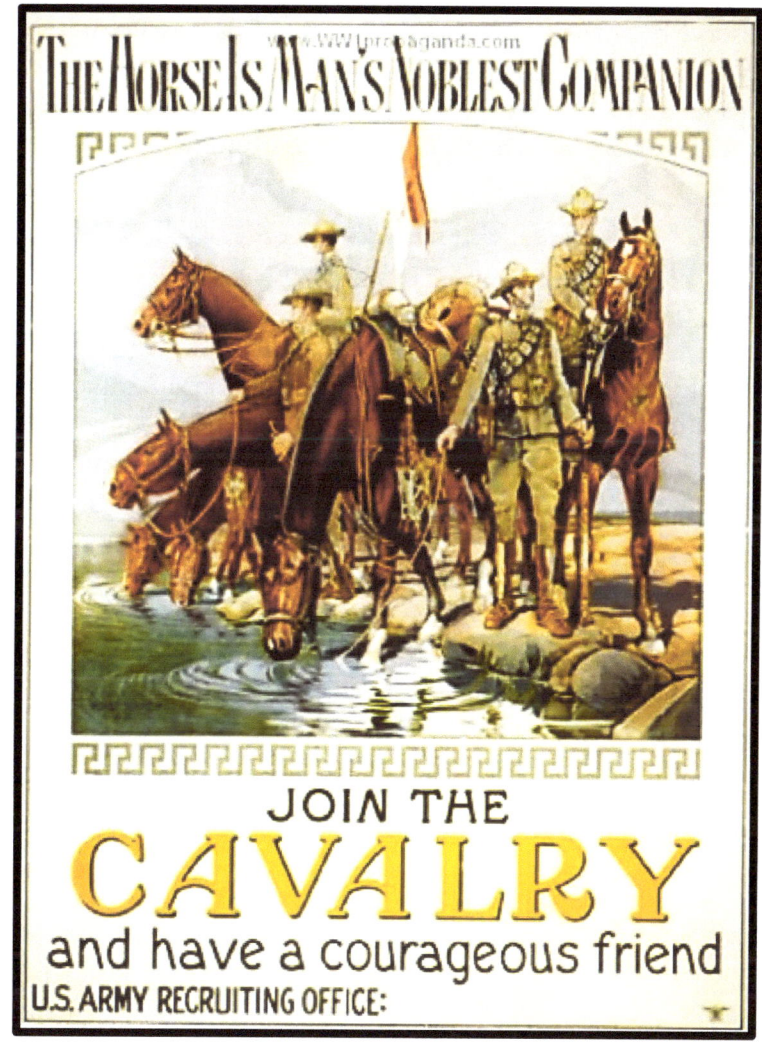

The Unsung Heroes of World War I

Allies' Cavalries

British Cavalry: In 1914, before the start of the First World War, just over 15,000 cavalrymen were serving in 31 British Army cavalry regiments. There were 3 Household Cavalry regiments and 28 line cavalry regiments consisting of 7 dragoon guards, 3 dragoons, 12 hussar, and 6 lancer regiments. In the British Army, the term "cavalry" was only used for regular army units. The other mounted regiments in the army which were part of the Territorial Force reserve were the 55 yeomanry regiments and 3 special reserve regiments of the horse. Several of the cavalry regiments, amounting to 6,000 men, was serving overseas in British India, South Africa, and Egypt. 12 of the regiments based in Great Britain were assigned to 4 cavalry brigades, 3 regiments per brigade, identified as being part of the cavalry division.
(cited from https://en.wikipedia.org/wiki/British_cavalry_during_the_First_World_War)

Canadian Calvary: When the war began, Lord Strathcona's Horse, a Canadian cavalry regiment, was mobilized and sent to England for training. The regiment served as infantry in French trenches during 1915 and was not returned to their mounted status until February 16, 1916. In defense of the Somme front in March 1917, Canadian mounted troops saw action.

Indian Cavalry: The Indian cavalry participated in actions on both the Western and Palestinian fronts throughout the war. Members of the 1st and 2nd Indian Cavalry Divisions were active on the Western Front, including in the German retreat to the Hindenburg Line and at the Battle of Cambrai.

Russian Cavalry: The Russians had 38 cavalry divisions. The myth of the presence of millions of fearsome Russian Cossacks has been called the great "Russian Rumor." There were indeed Cossacks among the Russian cavalry but certainly not millions. The Germans made a move to retreat their troops southeast as it neared Paris to avoid encountering the supposed millions of Cossacks. The moving of German troops gave the Allies time to plan well their advance at the Battle of the Marne by mid-September. The "Russian Rumor" played a significant part in leading on the Germans to a very major tactical military error. Subsequently, the scheming reports gave way to the victory of the Allies. Senior military officials have recognized it as a factor in the successful war efforts that followed. *(cited from https://www.warhistoryonline.com/war-articles/researcher-debunks-Russian-rumor-world-war-cossacks.html)*

At the outset of the war a troop of crusader knights–in full Medieval armor–marched right up to the governor's house in the Georgian capital, then called Tiflis (modern-day Tbilisi). "Where's the war?" They asked. "We hear there's a war." The Knights were known locally as Khevsurs, a group of fighters allegedly descended from Medieval Crusaders, whose armor bore the motto of the Crusaders, as well as the Crusader Cross (which now adorns the flag of the modern Republic of Georgia). They fought alongside the Russian army. *(cited from http://www.wearethemighty.com/articles/these-crusader-knights-answered-the-call-to-fight-world-war-i)*

French Cavalry: At the outbreak of the war the French army retained colorful traditional uniforms of the 19th century. These included conspicuous features such as blue coats and red trousers. The French cuirassiers (armored cavalry units of the French army) wore plumed helmets and breastplates almost unchanged from the Napoleonic periods. From 1903 on several attempts were made to introduce a more practical field dress, but these had been opposed.

The Unsung Heroes of World War I

ANZAC (Australian and New Zealand Army Corps) Cavalry: Initially Australia promised 4 regiments of Light Horse, 2000 men, to fight in the British cause. By the end of the war, 16 regiments would be in action. The Light Horse was seen as the "national arm of Australia's defense" and young men, most from the country, flocked to join. Many brought their own horses, and some even brought their dogs. It all seemed like a great adventure. A wonderful movie was made about Australia's horses in the war; "The Waler: Australia's Great War Horse." The Germans said of the ANZAC, "Anzacs and horses will go where no man will go. They are madness."

Italian Calvary: In the first decades of the 20th century the Italian Cavalry School at Tor di Quinto near Rome was–along with the French Cavalry School at Saumur–the leading institution for horsemanship in the world. Thanks to the observations and training by Federico Caprilli the Italian Cavalry School completely shaped their style around the horse. He studied how horses jumped without a rider and adapted to it rather than the opposite. The Italian front was either mountainous terrain or trenches, and thus horses were not used to a large extent on the Italian front. The Italian army primarily used horses for transport but struggled and sometimes failed to supply the troops sufficiently in the tough terrain.

Central Powers' Cavalries

Ottoman/Turkish Cavalry: The Ottoman cavalry mainly participated in surveillance, distracting the advancing enemy, closing gaps in army formations, and chasing the retreating army. The cavalry made up about 15% of the Ottoman army. There were three kinds of cavalry: light cavalry, field cavalry, and heavy cavalry.

Austro-Hungarian Cavalry: The cavalry was the most traditionalist and conservative arm of the Austro-Hungarian armed forces, some regiments tracing their lineage back as far as the Thirty Years' War of the 17th century. Like the French, they maintained their outdated, colorful uniforms at the beginning of the war. As the war continued, it became difficult to supply remounts to keep up with the casualties among the horses, and each year a larger proportion of the cavalry was dismounted to serve as infantry in the trenches.

German Cavalry: The history of the German Cavalry in World War I is one of decline. In September 1916, the establishment of cavalry regiments within the Cavalry Divisions was reduced to 675 horses instead of 769. The Supreme Command did not stop there, but also took away the horses of entire regiments and used them as infantry. Manfred von Richthofen (photo left) aka The Red Baron (1892-1918) earned widespread fame as a World War I ace fighter pilot. After starting the war as a German cavalry officer on the Eastern Front, Richthofen served in the infantry before getting his pilot's license.

Horses and Mules on the Fronts

Over 1,000,000 cavalry horses were kept on all fronts during the war. Many cavalry generals dreamt of heroic cavalry charges. Great numbers of men and mounts were "kept in the wings" waiting for the chance–some were converted to infantry units and horses were used for transport and other needs. The "Great War" ended the cavalry charge and the use of swords in the field. Modern ranged weapons beat close contact combat almost every time. In the right circumstances, however, the cavalry was able to "save the day" even in modern war. Despite the instinct of "fright–flight" horses and mules served bravely and valiantly throughout the war.

The Western Front

The contribution of cavalry to the fighting on the Western Front has been consistently underestimated by historians. Through an analysis of the performance of mounted units in these battles, using data principally obtained from the unit War Diaries, as well as other primary sources, it is argued that cavalry were both much more heavily involved in fighting on the Western Front, and more effective, than has previously been acknowledged. *(cited from a thesis by David Kenyon, Cranfield University on the topic of cavalry in World War I)*

In the very early days of the war, cavalry was a devastating weapon when used against infantry. A British cavalry charge at the **Battle of Mons**, August 23, 1914, (one of the battles of the Battle of the Frontiers) was enough to hold off the advancing Germans. Despite being outnumbered by the Germans 3 to 1 and having to retreat *(pictured left)* it still ended as a success for the British army. They achieved their objective, to prevent the French Fifth Army from being outflanked and that was a big moral victory for the BEF.

The war became a stalemate when the Allies won the **Battle of the Marne**, September 5-12, 1914. It was the second major clash on the Western Front and one of the most important battles of the war. The German retreat left the Schlieffen plan in ruins and dashed German hopes for a quick victory in the west. Its army was left to fight a long war on two fronts. While radio intercepts and aerial reconnaissance used in the battle presaged the future of warfare, echoes of the past remained in the cavalry troops charging on horseback, soldiers in red pantaloons charging behind commanders with swords drawn and drummers providing a musical soundtrack to the battle. *(cited from https://www.history.com/news/the-first-battle-of-the-marne-100-years-ago)*

Over 2,000,000 men fought in the First Battle of the Marne. More than 500,000 were killed or wounded. Both sides dig in, and trench warfare begins.

French 75 mm field gun with aircraft in the background during the Battle of the Marne

The Unsung Heroes of World War I

One of the last cavalry charges of the war came at the **Battle of the Somme** in 1916. The attack was on July 14th on High Wood–a German strongpoint that was holding up the British advance. Men from the 20th Deccan Horse *(pictured right),* an Indian cavalry unit, attacked the German positions. Armed with lances and despite going uphill which slowed down the charging horses, some men reached the woods. Some Germans surrendered when confronted by cavalry in woodland–something they could not have expected. However, the attack, while brave, was very costly with 102 men killed along with 130 horses.
(cited from http://www.historylearningsite.co.uk/world-war-one/the-western-front-in-world-war-one/cavalry-and-world-war-one/)

The **Battle of Verdun** (February 21 - December 18, 1916) was the longest battle in the First World War, lasting for almost a year. The German Chief of Staff, General von Falkenhayn, decided to attack Verdun, which had become a symbol of defiance and national pride to the French people. The stand made by the French over the following months into the autumn and winter of 1916 became legendary. *(left: French reserves and their horses resting in a river on their way to Verdun. "They shall not pass" is a phrase which will always be associated with the heroic defense of Verdun.)* The battlefield became known as "the mincing machine" to the French troops, but they continued to hold out, and Verdun was never captured.

The Unsung Heroes of World War I

The Eastern Front

Winston Churchill called the Eastern Front "The Unknown War." The Eastern Front saw bloody and chaotic clashes with cavalry officers leading suicidal mounted charges against dug-in machine-guns and modern artillery. As Winston Churchill famously noted, "In the west, the armies were too big for the land; in the east, the land was too big for the armies."

The Cossack cavalry of Plehve's Fifth Army discovered the gap left between the Austrian First and Fourth armies by the redeployment. General Nikolai Ivanov, commander of the Russian Southwestern Front, immediately sent his Fifth Army in, supported now by the newly arrived Ninth Army. Before Conrad could regroup his forces, the Russians had defeated the Austrians soundly at **Rava-Ruska** (September 3, 1914) and were advancing on both Przmyśl and the vital passes of the Carpathian Mountains that guarded the Hungarian plain. The Austro-Hungarian Army never truly recovered. It had lost over 300,000 men—nearly a third of the effective force — and a good percentage of its officer corps in the offensive. The battle was part of the series of engagements known as Battle of Galicia. *(cited from https://encyclopedia.1914-1918-online.net/article/eastern_front)*

Italian Front

The rugged terrain of the Italian Front made the use of cavalry difficult. In many instances the Austrian forces occupied higher ground, *(photos: The Austro-Hungarian supply line over the Vršič pass, October 1917)* and even though the Italians outnumbered them 3 to 1 at the battle of **Isonzo in 1915, the Italians were not able to gain any ground.** Like most militaries, the Italian army primarily used horses for transport but struggled and sometimes failed to supply the troops sufficiently in the tough terrain.

The Unsung Heroes of World War I

Gallipoli

(left: Allied troops land at Gallipoli and prepare their supply carts, 1915).

The Allies did not have good maps of the landing beaches or good intelligence on the disposition of the Turkish defensive positions. They never made it much past the beach. The withdrawal was about the only thing that went well. The New Zealand and Australian Division had two mounted brigades assigned to it. Some draught horses accompanied the divisional artillery and transport and supply units to Gallipoli in April 1915 to assist with their work. *(left: Allied troops land at Gallipoli and prepare their supply carts, 1915).* But the conditions proved unsuitable for horses. Some of those landed remained, but many were not landed or were soon evacuated and returned to Egypt. When the New Zealand Mounted Rifles Brigade was sent to Gallipoli in May 1915, it was as infantry. Their riding horses remained in Egypt. Mules and donkeys coped better with the shortage of water and the steep terrain; they were used to transport supplies such as water and ammunition by cart and on their backs. When the men returned to Egypt, they valiantly tried to find their horses among the thousands that were left there waiting for their return. *(cited from https://nzhistory.govt.nz/war/nz-first-world-war-horses/egypt-gallipoli)*

The Middle East

Conditions in the Middle East allowed for the cavalry to be a decisive factor in winning the war. The Desert Mounted Corps consisted of 20,000 Australian, New Zealand, British and Indian cavalry and mounted infantry. Marches of 60 miles a day with little water in arid conditions were common. Walers were the type of horse used by some of the light horsemen in the campaign in the Middle East. The light horse combined the mobility of cavalry with the fighting skills of infantry. They fought dismounted, with rifles and bayonets. However, sometimes they charged on horseback, notably at Magdhaba and Beersheba. The horses were called Walers because, although they came from all parts of Australia, they were originally sold through New South Wales. The **Battle of Beersheba** *(right)* was one of the greatest cavalry battles in history. Australian light horse troops had been marching for 24 hours and desperately needed water. The town of Beersheba had to be taken at all costs to gain access to the wells. Orders came to charge the town at full gallop armed only with bayonets. The Turks were so stunned that they forgot to lower their sights on the guns and the horses easily jumped both lines of trenches. Australians took the town in less than 10 minutes and took 2,000 prisoners. The victory was the beginning of taking Palestine from the Turks and Germans. Modern Israel would have never been created if it were not for the capture of this town and the success of the entire Palestine campaign.

Lawrence of Arabia (T.E. Lawrence), a British officer, was sent to meet Amir Feisal whose tribesmen had been attempting to besiege Medina. Lawrence was able to unite Arab tribes into guerilla warfare that proved to be very successful in the Middle East.

The Unsung Heroes of World War I

The **Battle of Rafa** fought on January 9, 1917, was the third and final battle to complete the recapture of the Sinai Peninsula by British forces during the Sinai and Palestine Campaign. This photo *(right)* was taken by Lieutenant Colonel Guy Powels as he and the ANZACs charged the Turkish defended town of Rafa. The photo is a graphic image of an actual mounted charge. The New Zealand troopers are riding through the clouds of desert dust towards the Turkish fortifications.

Two British officers posing in front of the Sphinx

Corporal Henry (Harry) Holt demonstrates a firing position beside his obedient horse, Jock. Men trained their horses to lie down while in battle to prevent them from being shot.

Sometimes a horse provided the only shad

Gloria Austin and Mary Chris Foxworthy

Care of Horses and Mules

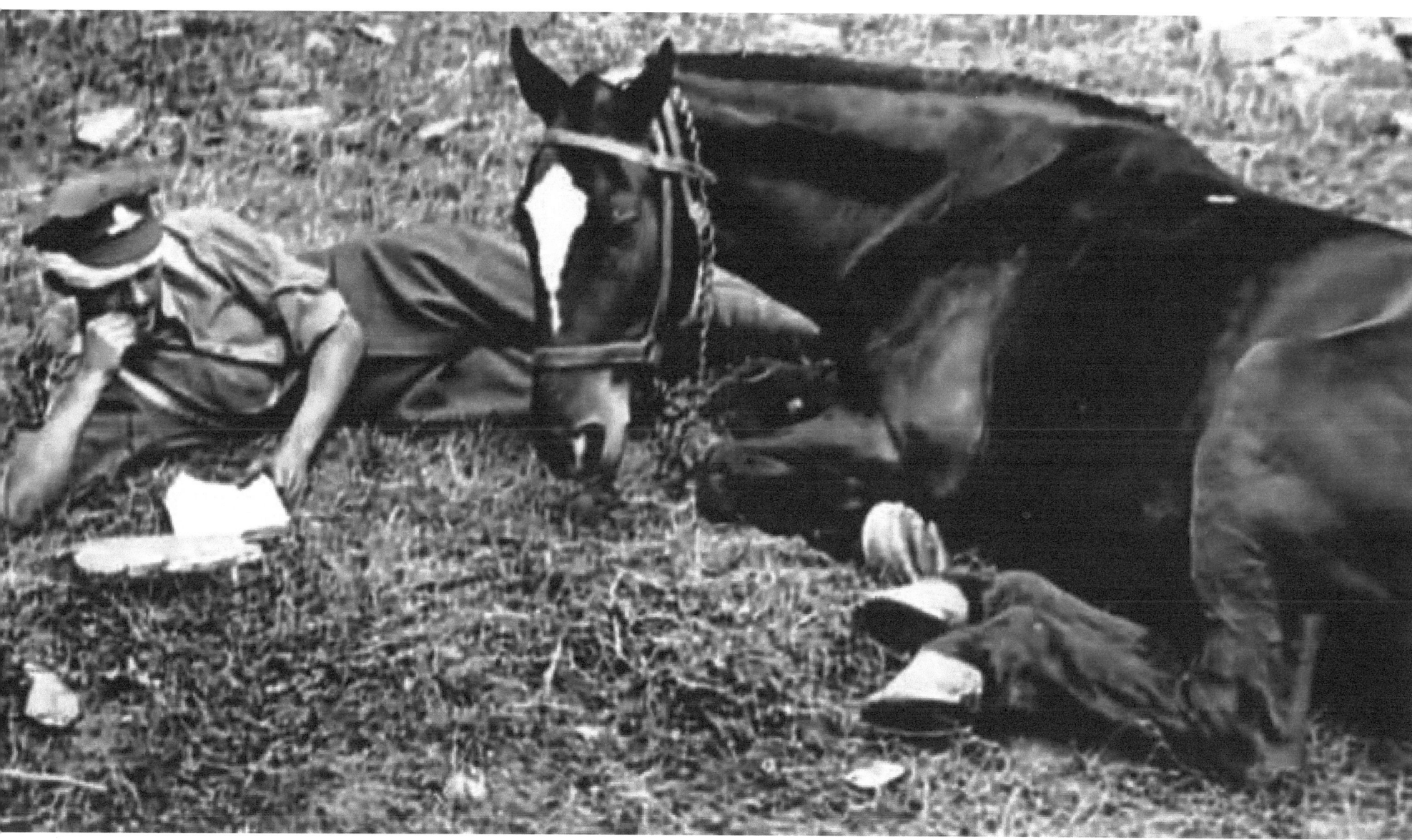

The Unsung Heroes of World War I

"I believe that every soldier who has anything to do with horse or mule has come to love them for what they are and the grand work they have done and are doing in and out of the death zones" *Captain Sidney Galtrey.*

Whether in the Allied Forces or Central Powers, the horse was the friend, companion, and hero to the men. Horses know no nationality. The German etching *(left)* says it so well, "I had a comrade." Much is written about the profound relationship between men and their horses. Life was more bearable for the men because of their equine companions. The horse often gave them something to focus on other than their own hardships. So close was the relationship with the men and their horses that when it came time to humanely destroy a horse, injured on the battlefield, most men were unable to do it. This duty fell to the officers. It was so important that this is done quickly and humanely that all officers were taught how to do this and it was listed in Chapter 7–section 33 of the Field Service Pocketbook, "To shoot a horse–lift up the forelock and place it under browband. Place muzzle of revolver almost touching the skin where the lowest hairs of the forelock grow." However, this was only done when it was determined the horse could not be saved. The health and care of the horse were paramount. For instance, in the deep mud of Passchendaele, men struggled with the horses and mules for over 12 hours to free them from the mud.

This painting "Friend" *(right)*, by the artist Heinrich Schubert, of the wounded Austro-Hungarian Hussar and his faithful horse sentimentally shows the relationship between horse and soldier.

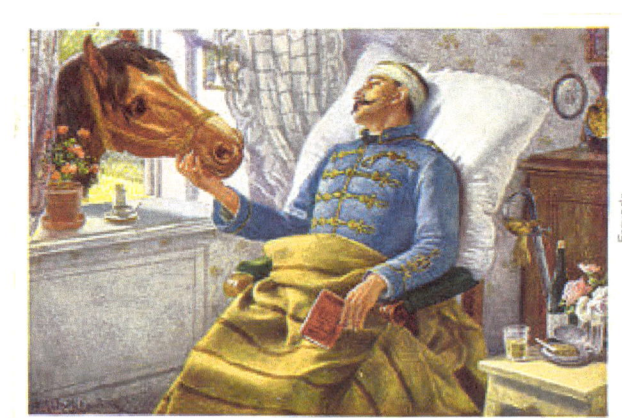

As was already noted in the section on obtaining horses, many horses were needed; it was easier to rehabilitate a horse already obtained and trained, therefore, the work of caring for the horses was extremely important. The work of veterinarians and the complete system for transporting and caring for horses was quite efficient. However, those handling the horses at the front often were "city boys" who had no knowledge of the treatment or care of horses. In 1915, Our Dumb Friends League put together a wonderful little 15-page booklet, "The Drivers', Gunners' and Mounted Soldiers' Handbook to Management and Care of Horses and Harness." You can read that little booklet here: *https://collection.nam.ac.uk/detail.php?acc=1994-06-217-8*

The Blue Cross Committee also became involved in caring for the horses. They decided to write to Field Marshal Earl Kitchener, Secretary of State for War, putting its services at the disposal of the authorities. Among the practical suggestions made were the provision of fully equipped and staffed horse hospitals and of a voluntary veterinary corps to supplement the military veterinary department. Their well-meaning offer was, however, politely turned down by the Army. In a letter from the War Office, signed by B. Cubitt, the Army Council said they fully appreciated " .....the value of the good work being done by your Society, but as ample provision has already been made for the care of sick and wounded horses at the Front, the Council regret that they cannot see their way to accept your very kind offer." Determined not to be put off, the Blue Cross Committee continued to supply materials and comforts for the horses to individual regiments throughout the war, and these were highly prized and appreciated by the men who had the horses in their care. Every annual report of the Blue Cross contained examples of letters received from the Front. Recognizing that horses have no nationality and that there were more ways of helping them than via the British Army, the Blue Cross sent an envoy to France with offers of assistance to the French Army. The reception there was rather more enthusiastic, and the Minister of War, M. M.A. Millerand, wrote: "I will add that instructions will be given to give this Society every facility for the organization of depots behind the firing line, where horses will be trusted to its care... When you inform the Society of the Blue Cross of these arrangements, will you also kindly express the lively gratitude of the French Government for the offer of its precious help in the work undertaken to cure the horses which have already rendered such great services, and in this way to reconstruct a material which represents one of the principal elements of strength of the armies." *(cited from https://www.bluecross.org.uk/blue-cross-war-book)*

Out of necessity, new preventative measures and procedures for the treatment of disease, infections, and wounds were developed by the Veterinary Corps in WWI. Even a table for operations was developed during WWI. Captain W. W. Long RCVS, Staff Sergeant Major H. Hayes and Staff Sergeant S. Nunn of the Indian Army Veterinary Corps in 1916 designed an operating table. A total of 4 tables were built. They received congratulations and received special promotions for their work.

The Unsung Heroes of World War I

The U.S. Army Veterinary Corps was established June 3, 1916, with the National Defense Act. At the beginning of World War I there were 72 veterinary officers and no enlisted men. Within 18 months this grew to 2,312 officers and 16,391 enlisted personnel. This rapid growth in the middle of the war required the establishment of recruitment, training, policies, procedures, and equipment.

The American Expeditionary Force (A.E.F.) required large numbers of animals to accomplish a variety of missions ranging from cavalry mounts, artillery transport to logistical supply and ambulance service. The rugged and muddy French terrain was better suited to animals than gas-powered engines. General Hagood, Chief of Staff of the Services of Supply, placed the number of horses and mules in the A.E.F. at 165,366 as of October 30, 1918. Disease detection, prevention, and treatment played an important part of the Veterinary Corps officer work. *(cited from http://veterinarycorps.amedd.army.mil/history/ww1/ww1.htm)*

The British had an excellent system of caring for the horses starting at the reception point for horses, then at the front, in the hospitals, and during the rehabilitation before return to the front. Much of the information on this topic in this section is again from the book, "The Horse and the War" written by Captain Sidney Galtery, a British Remount Service officer.

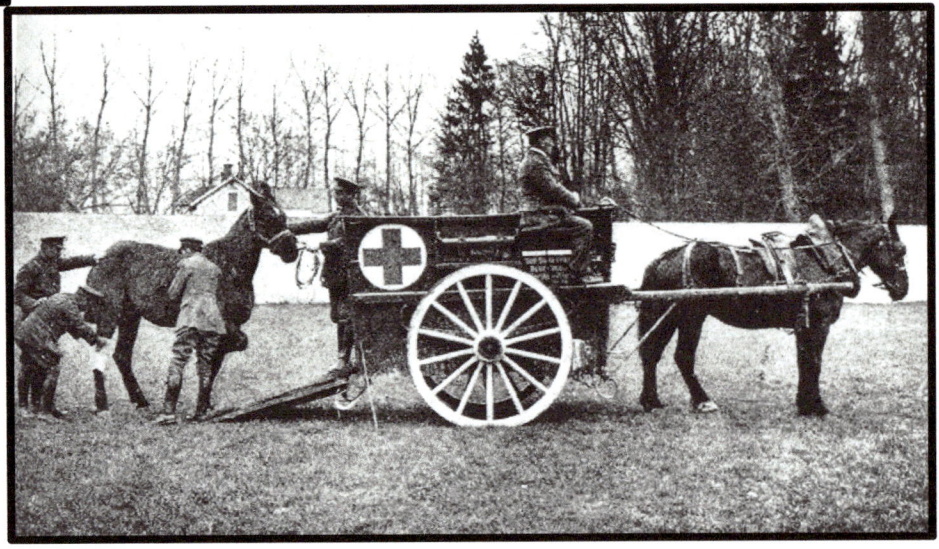

It was also important that those caring for the horses were properly trained to do so. "It was about this time that the Commander-in-Chief showed his watchfulness and zeal for the welfare of his horses; and one outcome, which I feel sure has had most excellent results, was the appointment to each corps of a chief horse master, who had under him subordinate horse-masters, each attached to minor units. They were ostensibly what their designation implies — experts in horse and stable management; and it has been their duty ever since to watch those units employing horses and to give useful advice for the improvement of the necessary hard lot of horses and mules on active service close behind the Line. Really efficient and tactful horse-masters have, I am sure, done good, though the splendid condition of the animals in France to-day has been primarily due to the better and milder winter. Then, the Director of the Veterinary Service in France has abundantly aided the good work by instituting at each of his hospitals a ten-day course of lectures and instruction for artillery and infantry transport officers. In this way, 50 officers and 300 N.C.O.'s have taken the course each month." *(cited from "The Horse and the War, p78)*

Grooming the horses was not only necessary but a welcome opportunity to spend time with a close companion. There are many stories of soldiers naming horses after wives and girlfriends at home so that they could talk to the horse as if they were talking to that person. In muddy conditions, it could take up to 12 hours to clean horses and their harnesses or saddles. Good grooming meant the horse was always ready for a battle at a moment's notice. Grooming also provided the opportunity to inspect horses for pain, wounds, skin conditions and sickness on a daily basis.

The Unsung Heroes of World War I

Food was a major issue. Rations for each horse included 12 pounds of oats, 10 pounds of hay, and some bran every week. Multiply that by a million, and you get problems with producing enough feed not to mention transporting it to the horses who were spread out across Europe and under near-constant enemy fire. The average ration of a supply horse was 20 pounds of fodder which was a fifth less than recommended. This meant the average battalion needed at least 7,480 pounds of oats and hay a week to feed its 56 horses. Gun horses were bigger and pulled heavier loads so required 30 pounds of fodder. They could spend up to 5 hours a day eating.

Often, the horses went hungry, and many also went without sufficient water. The British army provided 2,978,301 tons of oats and 2,460,301 tons of pressed hay as fodder during the war. *(cited from http://www.todayifoundout.com/index.php/2014/03/horses-world-war/ and http://www.bbc.co.uk/guides/zp6bjxs)*

The shelter was another problem. Most horses were simply attached to a picket line without a roof over their heads, and some stood in trenches. They were subjected to cold, wind, snow and rain. Horses were a constant target of the enemy. Efforts to poison their feed and bomb the hospitals and picket lines were a constant threat. Guards kept watch over horses and hospitals were often camouflaged. *(picture below)*

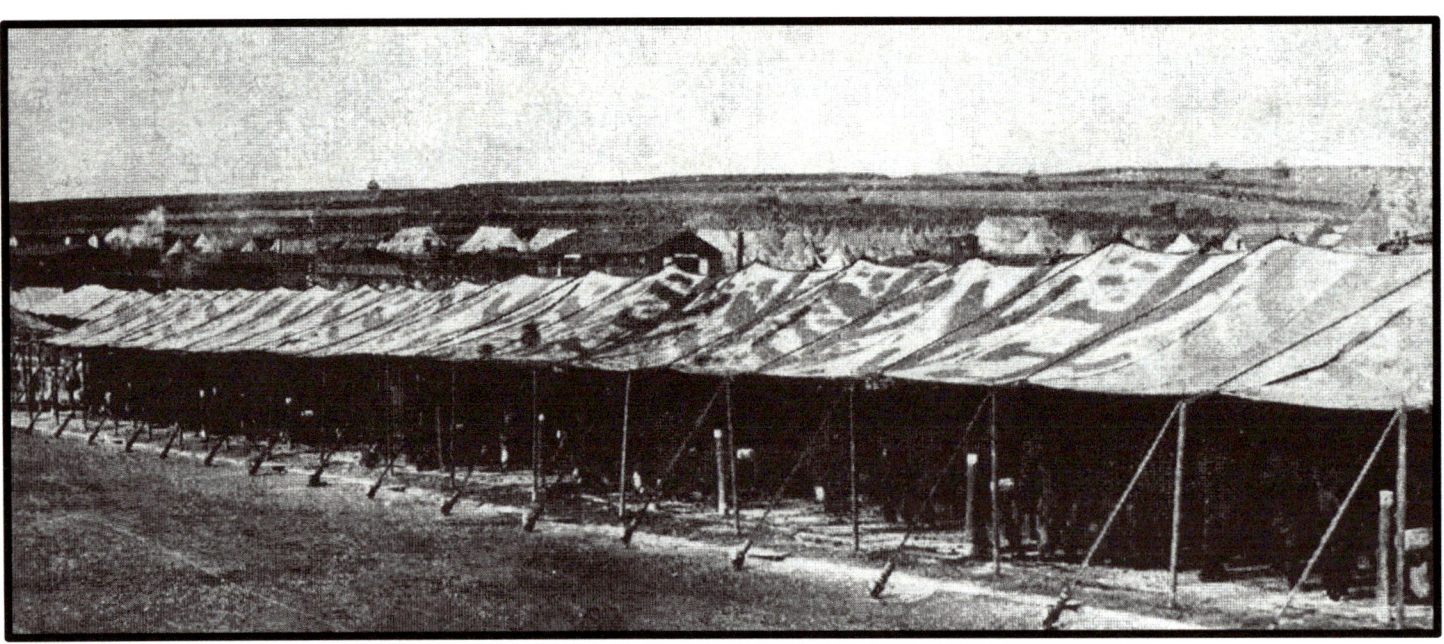

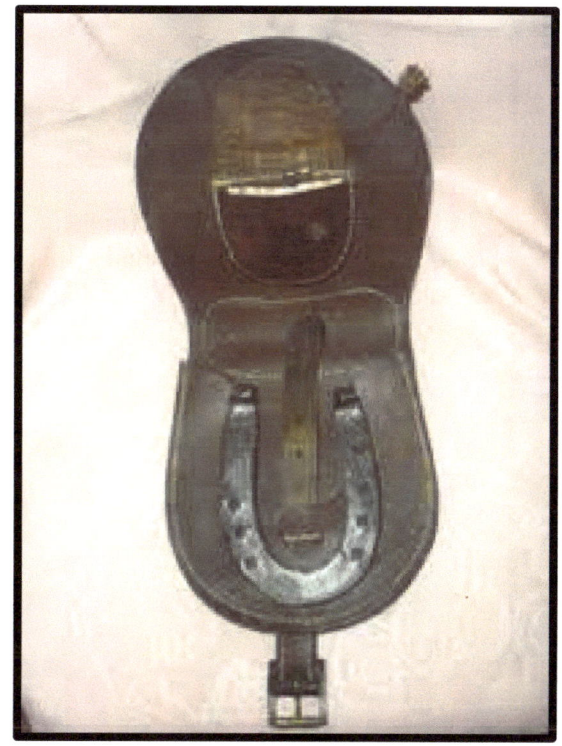

Farrier care was incredibly important and constant. However, if work was needed immediately, all horses carried a horseshoe pouch. This leather case *(left)* was designed to carry two spare horseshoes and necessary nails.

Despite the best intentions and efforts of the men at the front, horses did experience disease and injury. It was Captain Galtrey's job to visit and supervise the care of the many thousands of horses and mules at the time of their arrival, on the front, and in hospitals in France. When visiting the front, he was always amazed at the excellent condition of the horses. The real enemies were the hard weather, the hard conditions under which horses worked and existed, and those diseases which were incidental to the collection and movements of horses and mules in great numbers. Horses that showed signs of debility or exhaustion had been sent to base hospitals or convalescent horse depots. The rate of healthy horses to sick horses in 1918 was 9 to1. The sick or debilitated horses were not kept at the front because that would jeopardize their recovery and perhaps cause other horses to become sick. Furthermore, long recovery for a horse was a doubtful proposition from a financial point of view.

The Unsung Heroes of World War I

Every endeavor was made for the comfort and safety of the horses. The Royal Society for the Prevention of Cruelty to Animals made grants of 100,000 pounds (that would be $2,497,594.03 in 2018 US dollars) to the cause of caring for sick and wounded animals. The money helped provide the accommodations, horse ambulances, and laboratory appliances. *(left: veterinary hospital in France)*

The saving of men's lives in movements to and from the fronts was largely dependent on the health and fitness of the horses and the saving of horse's lives.

Mange was the single most cause of sickness and death. If caught on time, the treatment for a cure was about 2 months. It attacked horses more than mules. It was caused by constant exposure to mud and the elements. Many treatments for mange were developed during the war including a gas treatment (left) and a dip bath (right)–which proved to be most effective. At the onset of the war, clipping winter coats was determined to be the best way to prevent mange. However, in the terrible winter conditions, horses became ill and depilated without their winter coats. Complete clipping of coats on horses and mules "at the beginning of winter is both folly and cruelty since it must deprive them of the warmth provided by Nature. They do say that the losses of the winter and spring 1916-17 were assisted by the clipping which was general, and the laws of logic and nature would seem to confirm the theory. But it is a point on which the expert and the veterinary specialists do not quite agree, and therefore there has been something of a compromise during the 1917-18 winter with certainly vastly improved results. The point made by the Veterinary Service, however, is quite intelligible. They say that the growth of a long coat hides mange and other serious skin troubles until it is too late, when eventually detected, to effect a speedy cure. Remount officers and others say that total clipping must cause a great wastage from debility and death and that it is better to clip, if at all, in the late autumn or very early winter. 1 am sure the veterinary officers agree that it is undesirable to deprive animals of their winter coats. It, therefore, becomes a question of arriving at the lesser of two evils, and I am sure the compromise of the fourth winter of the war has been the right and sane one." (cited from, "The Horse and the War," p.88)

Caltrops (left) are ancient war weapons that were a great danger to horses. The Germans often scattered the ground with these as they were retreating. The caltrop would sink in the mud and not be visible. If a horse stepped on one, it was often fatal. They are still used in warfare today causing damage to vehicles with pneumatic tires and people in soft-soled shoes/boots.

White horses were not often used, and the few that were often were camouflaged (right) to protect them and the men with them.

The Unsung Heroes of World War I

"Gunshot injuries naturally fluctuate according to what is going on at the front. Veterinary surgeons do indeed owe much to the experience war had brought them. Especially is this so of those who, day after day, have been engaged in the operating theatres searching for and extracting the cruel jagged shell splinters, shrapnel bullets, and bomb splinters, and in many different ways bringing relief to the poor suffering creatures. Theirs has, indeed, been humane and splendid work, for by their skill and knowledge they had saved the lives of thousands that would have been doomed in days gone by when surgery was nothing like as advanced as it is today. Then, too, their operations have been assisted by the aseptic methods of sterilization of wounds and instruments in place of the antiseptic methods once favored. *(right: operation in a Veterinary Hospital on a horse under chloroform).* One officer I know is very proud of twenty-three pieces of shrapnel of all sizes which he extracted from one horse. That same horse was in due time restored to active service. When I think of the great work being done by the Veterinary Service, of the immense strides it has made forward in research, surgery, and the study of disease, I wonder that the authorities have not established schools on the spot for the training of those who must one day fill the ranks." *(cited from: "The Horse and the War," p.103)*

In the spring of 1918, there were 30,000 horses and mules in veterinary hospitals and convalescent depots. 551,960 horses and mules were admitted to veterinary hospitals and convalescent depots from the beginning of the war until mid-February of 1918. 394,768 (about 71.5%) were passed out as cured, 34,327 still under treatment, 16,215 died, 106,650 were destroyed or cast and sold. This was in France only. *(cited from "The Horse and the War," p104)*

The various veterinary corps kept records of the diseases, injuries, and losses that befell the horses and mules. Sadly, not all horses and mules made it to depots and hospitals to be saved, and some did not survive once they were at the hospitals. The figures (below) from the US Veterinary Corps were fairly consistent throughout the war among all the veterinary corps. Losses of horses and mules were attributed to Mange 47%, Influenza 4.03%, Lymphangitis .03%, Glanders .17%, Wounds and lameness 2.06%, Mustard Gas .03%, Debility 7.05%, Misc. 1.75% *(cited from http://veterinarycorps.amedd.army.mil/history/ww1/ww1.htm)*

"The horses that could not be cured or saved were 'cast.' The Veterinary Service is the chief casting authority; only on the score of age and unsuitability for any specific job in the Army can Remount Authorities exercise the veto of casting. Since a trifle over 20% of the horses admitted to hospitals never return to active service, it will be understood that castings are on a big scale. Do not for a moment imagine that this 20%, represents a dead loss. From (a) sales to farmers for work on the land; (b) sales to horse butchers for food; and (c) reduction of carcasses of any animals not suitable as food for by-products of skin, fat, bones, flesh, hoofs, etc., the average sum of 50,000 pounds a month for the British government so the Veterinary Service is using its splendid organization to get the utmost possible out of our war animals whether in life or in death." *(cited from: "The Horse and the War," pp 108-109)*

Both illustrations by: Captain Lionel Edwards from the book "The Horse and the War"

left: cast horses on the road to the place of sale *right: the start for the "casters" new home*

The Unsung Heroes of World War I

Real Stories of Horse and Mule Heroes

"Put your life on them, and they'll get you home."

THE "OLD BLACKS"

A gun team of six beautiful black horses who survived the whole war was chosen to pull the carriage of the Unknown Soldier to mark the Armistice in 1920. The team, affectionately known as "The Old Blacks," were retired in 1926.

In 1916, a Church of England clergyman serving at the Western Front in World War I spotted an inscription on an anonymous war grave which gave him an idea. The Reverend David Railton caught sight of the grave in a back garden at Armentieres in France in 1916, with a rough cross upon which was penciled the words "An Unknown British Soldier." In August 1920 Mr. Railton wrote to the Dean of Westminster, Herbert Ryle, to suggest having a nationally recognized grave for an unknown soldier. The idea was presented to the government and quickly taken up.

The unknown warrior's body was chosen from some British servicemen exhumed from four battle areas - the Aisne, the Somme, Arras, and Ypres. These remains were brought to the chapel at St. Pol on the night of November 7, 1920, where the officer in charge of troops in France and Flanders, Brigadier Gen L J Wyatt, went with a Colonel Gell. Neither had any idea where the bodies, laid on stretchers and covered by union jacks, were from. "The point was that it literally could have been anybody," says Mr. Charman. "It could have been an earl or a duke's son, or a laborer from South Africa. General Wyatt selected one body and the two officers placed it in a plain coffin and sealed it. The other bodies were reburied.

The next day the dead soldier began the journey to his final resting place. The coffin was taken to Boulogne and placed inside another coffin, made of oak from Hampton Court and sent over from England. Its plate bore the inscription: "A British Warrior who fell in the Great War 1914-1918 for King and Country". This second coffin had a 16th Century sword, taken from King George V's private collection, fixed on top. The body was then transported to Dover via the destroyer HMS Verdun and taken by train to London.

On the morning of November 11, 1920 - two years to the day after the war had ended, the body of the unknown warrior was drawn in a procession through London to the Cenotaph. This new war memorial on Whitehall was then unveiled by George V. At 11:00 there was a two-minute silence, and the body was then taken to nearby Westminster Abbey where it was buried, passing through a guard of honor of 100 holders of the Victoria Cross. In a particularly poignant gesture, the grave was filled with earth from the main French battlefields, and the black marble stone was Belgian. And at the exact time, Britain was interring its unknown warrior, France was doing the same - burying its Soldat Inconnu at the Arc de Triomphe in Paris. An estimated 1,250,000 people visited the Abbey to see the grave in only the first week. *(cited from http://www.bbc.com/news/magazine-11710660)*

The Unsung Heroes of World War I

Warrior

The Honorary PDSA Dickin Medal was presented posthumously in 2014 to famous war horse Warrior, dubbed "the horse the Germans could not kill."

The bravery of the thoroughbred was documented in a book written by General Seely, in 1934.

After arriving on the Western Front on August 11, 1914, with General Seely, Warrior stayed there throughout the war, surviving machine gun attacks and falling shells at the Battle of the Somme. He was dug out of the mud of Passchendaele and twice trapped under the burning beams of his stables, surviving many charges at the enemy and proving an inspiration to the soldiers he was fighting alongside. Despite suffering several injuries, Warrior survived and returned home to the Isle of Wight in 1918, where he lived with the Seely family until his death aged 33.

(Cited from https://www.telegraph.co.uk/history/world-war-one/11069681/Heroic-First-World-War-horse-Warrior-receives-animal-Victoria-Cross.html).

The horse was often a source of support and morale for the troops. They would often pat him and say, "good boy"–because of course you could not pat the General on the back and say that! In 1934 horse and rider celebrated a centenary ride–Warrior was 30, and General Seely was 70. *(right Warrior age 30 with General Seely and Queen Mary)*

Midnight

Midnight *(left)* was born and bred at Bloomfield Homestead in Scone, New South Wales, a stud which still produces quality horses today. The mare, a "Waler," took part in the last great cavalry charge in history. In October 1917, the Australian Light Horse were closing in on Beersheba. The water wells had been mined by the enemy and needed to be taken intact. Command decided on a great gamble. One of the horses taking part in the charge was Midnight ridden by Guy Hayden. Midnight jumped a trench and took a bullet intended for Guy. The bullet went into Hayden's back, through the saddle, and into the horse. The family still has the bullet. *(cited from http://www.abc.net.au/local/stories/2014/08/15/4067966.htm)*

Pie

Parts of the 10th Australian Light Horse Regiment were the first troops formally to enter Damascus. After a sleepless night, at 5 am on October 1, 1918, they began to move into the city, arriving in the city square before 7 am. The commanders entered the Serai (Town Hall), where the city administrators were meeting. Captain Arthur Olden on his Waler horse, Pie, demanded protection for his troops, and in return promised that they would not harm the populace. Governor Emir Said Abd el Kader, who had only been instated the night before, replied, "In the name of the civil population of Damascus, I welcome the British army." Olden then got this assurance formally in writing, then left to continue their ride through town to block the Homs road. This was made difficult by the exuberant crowds now eager to welcome the liberators. They then went in pursuit of the retreating Turks and Germans towards Aleppo, leaving others to take charge of the city. *(cited from http://theanzaccall.com.au/stories/damascus.html)*. Two hours later Lawrence of Arabia arrived in a Rolls Royce and accepted the formal surrender; there was a political need for Damascus to be seen to be liberated by the Hashemite army led by Feisal that had fought its way north from the Hejaz. Olden's family still has the handwritten note from Governor Emir Said Abd el Kader.

The Unsung Heroes of World War I

Bill the Bastard

Bill the Bastard was a 17.1 hand, fiery, cranky chestnut gelding. He is one of Australia's greatest war horses and became a legend, famed for his incredible stamina and for saving many soldiers' lives. An Australian-bred "Waler," Bill earned his not so illustrious nickname 'Bill the Bastard' because he was famous for his incredible bucks, unseating rider after rider after rider. But it was his relationship with Michael Shanahan, which gave the fierce chestnut the chance to become the hero he was meant to be. Shanahan and Bill were among the 100,000 horses who fought in the pivotal Battle of Romani, in the heat of the desert. With the Turks and Australians just 35 meters apart, the raging battle was fierce with the right flank under assault. Shanahan was able to rescue four injured soldiers on Bill, who stood his ground and didn't cut and run like some other horses. Once mounted and hanging on in the stirrups, Bill carried the soldiers for just over a kilometer to safety, all the while under heavy fire. *(painting below by Peter Smeeth)* Any other horse would most likely have collapsed, but Bill's stamina played an instrumental role, enabling him to continue for six hours straight. Shanahan was also injured in the battle and Bill, ever the loyal steed carried him many kilometers to safety. After his instrumental efforts in the Battle of Romani, Bill was officially retired from battle, but he would still help out as a pack horse from time to time, carrying machine guns or leading the line. He was seen as a symbol of strength to the troops. *(cited from http://www.globetrotting.com.au/bill-the-bastard/)*

Cupid

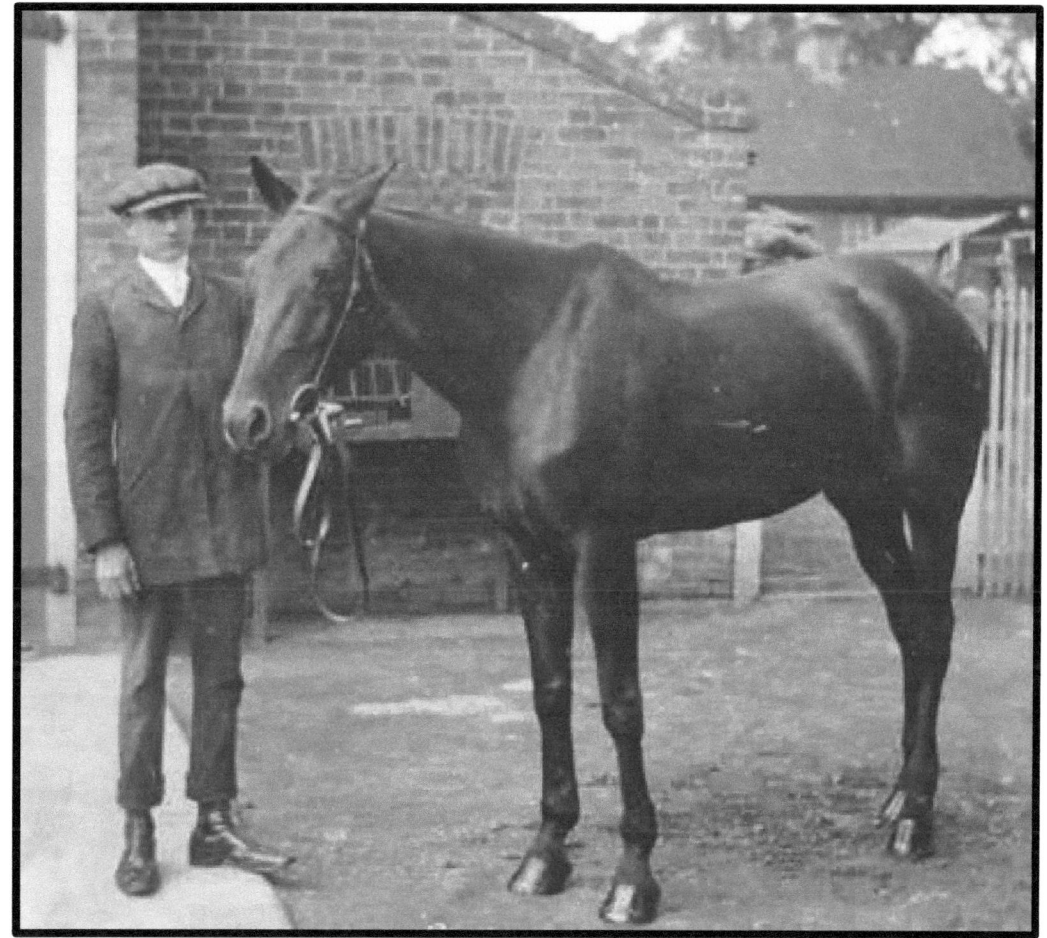

The real story began in 1911, when Ranald Laurie, a City broker, and Martin Laurie's great-grandfather, bought Cupid for his son as a present for his 15th birthday. Vernon and Cupid spent "three idyllic years" hunting together in the countryside around the family's farm in Essex. Ranald, who was a lieutenant colonel in the Territorial Army, never expected the war, and in fact resigned his commission in February 1914, because he had reached his 40s. However, war was declared just six months later, and Lt Col Laurie enlisted at once, joining the 271st Brigade of the Royal Field Artillery. He bought 131 horses for the battery and readied the family's horses for action. On December 2, Vernon finished school, and two days later he was on active service, as a second lieutenant. Both father and son began to put their horses through battle training. For more than three years, Cupid rode back and forth from the trenches in France, trekked across the Sinai desert and dodged shellfire during bloody battles for Gaza. By the end of the Palestine campaign, only 22 of the 60 horses they had brought from Essex survived, making it even more remarkable that all three horses who arrived in Egypt, as well as father and son, made it through to the end. Now, despite a Government ruling that no horses were to be shipped home, 2nd Lt Laurie harbored "faint hopes" that he could get Cupid back to Essex. It was not to be. He returned home first, in January 1919, and on February 24, his father wrote to tell him that he had to "destroy poor old Cupid" after she had "a frightful set to with a wandering mule at night" and was "horribly mauled." She had survived the war, only to be killed by friendly fire. "I think it is very sad," said Mr. Laurie. "It was hugely upsetting for my grandfather because he adored the horse." So much so that he instructed a veterinary sergeant to clean up one of Cupid's hoofs, which was shipped back to Britain and mounted in brass as a doorstop which Mr. Laurie still uses today, inscribed: Cupid 1909-1919. The story of Cupid is chronicled in the book, Cupid's War, by Martin Laurie. *(cited from https://www.telegraph.co.uk/history/world-war-one/11202449/The-real-life-War-Horse-Cupid-the-bay-mare-from-Essex.html)*

The Unsung Heroes of World War I

Sandy

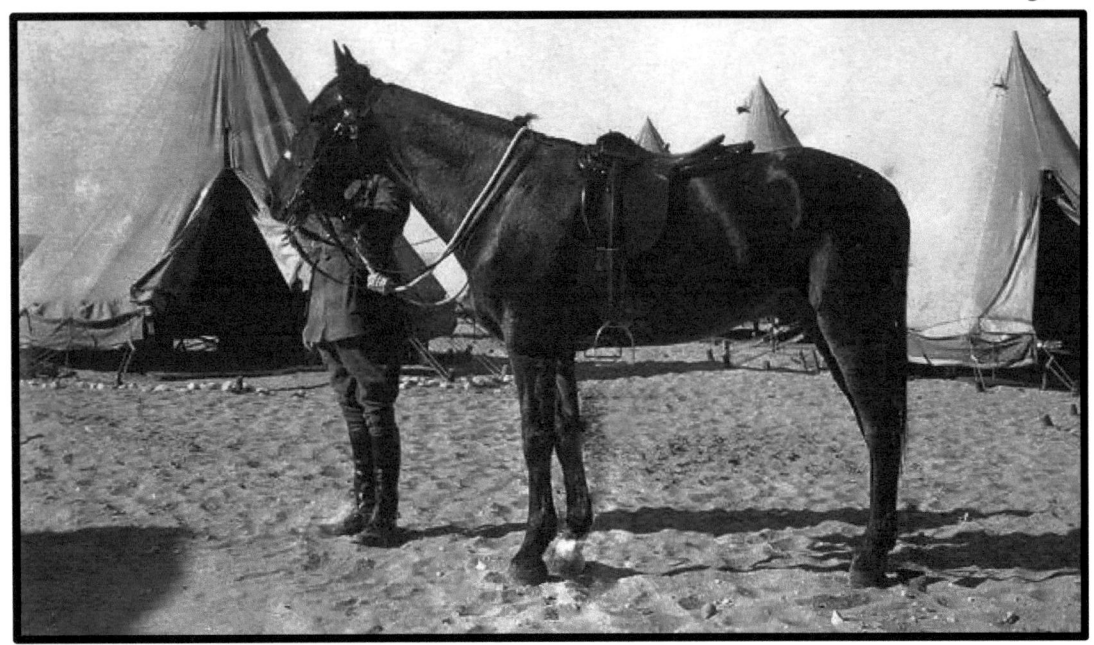

Sandy was the only horse from Australia to return from the First World War. 136,000 "Walers" (the general name applied to Australian horses abroad) were sent overseas for use by the Australian Imperial Force and the British and Indian governments. Sandy belonged to Major General Sir William Bridges, who was killed at Gallipoli. *(left: Sandy and Major General Sir William Bridges)* He was one of 6,100 horses who had embarked for Gallipoli. However, very few of the animals were put ashore, as Lieutenant General Sir William Birdwood decided there was no room or requirement on Anzac Cove. On May 5th Birdwood sought approval to send the horses back to Alexandria.

From August 1, 1915, Sandy was in the care of Captain Leslie Whitfield, an Australian Army Veterinary Corps officer in Egypt. Sandy remained in Egypt until he and Whitfield were transferred to France during March 1916. In October 1917 Senator George Pearce, Minister for Defense, called for Sandy to be returned to Australia for pasture at Duntroon. In May 1918 the horse was sent from the Australian Veterinary Hospital at Calais to the Remount Depot at Swaythling in England. He was accompanied by Private Archibald Jordon, who had been at the hospital since April 1917 and classed as permanently unfit for further active service. After three months of veterinary observation, Sandy was declared free of disease. In September 1918 he was boarded on the freighter Booral, sailing from Liverpool and arriving in Melbourne in November. Sandy was turned out to graze at the Central Remount Depot at Maribyrnong. Sandy saw out the rest of his days at the Remount Depot. *(cited from https://www.awm.gov.au/articles/encyclopedia/horses/sandy)*

David

Four officers clubbed together funds to purchase and then ensure a peaceful retirement for a horse called David. This extraordinary animal had served as a youngster in the Boer War, and as a veteran on the Western Front, every day without fail (except for the one occasion he was wounded). *(cited from https://www.dailypost.co.uk/news/local-news/horse-penmaenmawr-who-served-first-2665570)*

Jezebel

"Certainly, it was not a nice name to give a mare, but then Jezebel was not a nice mare. It sounds an exaggeration to say that one horse could upset a whole squadron, yet Jezebel came very close to doing that to A Squadron of the 8th Light Horse. She came to us at Heliopolis after the evacuation of Gallipoli, sent up from the Remount Depot to replace a horse with a broken leg. In the opinion of A Squadron, the Army buyer who picked Jezebel was either blind drunk or else his knowledge of horses had been gained from reading the Saturday evening race results. She was a tall, rangy brute with a wicked eye, a Roman nose, a long quivering under-lip and the sourest nature ever implanted in a horse.

 Trooper "Snow" Matson fell for her and Snow really deserved a better fate. He was a born horseman, a man who understood and loved horses. In civvy life, he'd been a station hand and drover, one of the long, lean types one sees everywhere in the Australian outback. On Gallipoli, he'd proved himself a good soldier, a cheerful hard case who could crack a joke in the most desperate situation. But no matter how he tried Snow couldn't work up a spark of affection for Jezebel. No man could. Grooming her was a hazardous, nerve-racking job from start to finish. When Snow ran the brush over her flanks, she lashed out with the vicious, raking kick of a soured mule; when he bent down in the region of her girth, he exposed the seat of his pants to a slashing bite. She kicked and bit the horse on either side of her just as impartially.

But it was on mounted parades that Jezebel really distinguished herself. Keep in line? Not her! She had a habit of snapping at the bit and bounding half a length ahead. When Snow dragged her back by main force, she'd run backward until a jab of the spurs sent her bounding ahead again. When she tired of that she'd swing sideways, bite the horse on her left and lash out at the one on her right.

At first, Snow tried kindness. "She's been bashed about; that's what's wrong with her," he reasoned. "They nearly all come good if you treat 'em right." Snow was that sort of bloke, a good patient horseman. But kindness was wasted on Jezebel. She took it as a sign of weakness. There was as much affection in her eye when Snow approached her with an extra ration of oats in her feed bag as there was when he changed his tactics and took to her with the flat of a shovel the day she ripped the back out of his one and only shirt.

Just before daylight one morning our scouts made contact with the enemy. Only a small force, we reckoned, for the rifle fire was light and spasmodic, and there was no shelling whatsoever. It was full daylight when we went into action, A Squadron leading in open formation. The rifle and machine-gun fire grew intense, and we got the order to trot. Ahead of us was a small sand ridge with a nice hollow on our side of it. The volume of fire increased, the trot turned into a canter and the canter to a mad gallop as we raced for the shelter of that hollow. We were "copping" it now! Horses and men were falling all along the line as we galloped through a hail of fire as hot as anything we'd struck on Gallipoli.

The Unsung Heroes of World War I

We reached the hollow and flung ourselves out of saddles in response to the shrilling of whistles and the shouting of orders. All, that is, with the exception of those who had fallen in advance and one man who didn't stop. We yelled, "Pull up, Snow! Come back! But Snow, tired out and weakened from twenty-four hours of almost unadulterated Jezebel, couldn't pull up. Jezebel had the bit in her teeth and was racing straight for the Turkish lines. To us, as we handed our horses over to the horse holders and threw ourselves flat on the crest of the ridge, it seemed incredible that a horse and rider could live through that murderous hail of fire, yet both came back untouched. Unable to stop her, Snow swung Jezebel in a wide arc and headed her back for her troop mates. She crashed into them, and Snow flung himself out of the saddle just as the horse holders started to lead them back out of range. 'Take her away,' Snow said wearily as he passed his reins to our section No. 3. "Take the sod away and shoot her." And then, with a flash of his old humor, 'I took her over to the Jackos and offered her to 'em for five 'disasters,' but they knocked 'er back. They wouldn't even waste a bullet on her'

Away from us, we could see the shells burst as the Turkish artillery searched the hollows for our led horses. My own horse was safe, I saw with relief as our section horse holder swung towards us; so also was Sweetheart, the little chestnut mare belonging to Charlie Conlon, my No.2. But of Jezebel, there was no sign. Catching sight of Snow, our horse holder yelled, "Jump up behind me, Snow. Jezebel's dead. 'Thank God for that,' the major said. 'There you are, Matson. I told you anything could happen in action. We'll be getting some remounts up after this. You can have your pick of 'em and see that you pick a good one this time.' 'My flamin' oath, I will!' Snow declared. 'Hey, major.' 'What, Matson?' 'D'you mind if I drop off 'ere? I just want to nick back and see 'er for meself – just to make sure she's dead.' 'You stay where you are,' the major commanded. 'If that damned mare turns up I'll see she never gets back into this squadron, even if I have to shoot her myself!' *(cited from http://www.lighthorse.org.au/resources/military-stories-ww1/a-story-about-a-horse-called-jezebel-in-my-grandfather2019s-squadron-in-ww1)*

Charlie

Charlie was a horse from Penmaenmawr owned by William Roberts, the local coal merchant. At the time dray horses were the main means of transportation, and Charlie was sent to the front in France to pull artillery guns. William's grandson Frank Roberts said: "The government commandeered as many horses as it could. Fortunately, Charlie survived the war, and he was transported back from France to Penmaenmawr. On arriving at Pen railway station, Charlie was released from his horse box, and without so much as a look around he walked out of the station yard, up to Paradise Road, across the main road and walked straight into his stable behind Fernbrook Road. He behaved as if nothing out of the ordinary had happened. The tale went down in our family folklore, my grandfather would tell the story of Charlie time and time again. He lived to a ripe old age and even worked when he was blind, delivering coal around the village." *(cited from https://www.dailypost.co.uk/news/local-news/horse-penmaenmawr-who-served-first-2665570)*

Kidron

Kidron was the war horse, ridden by General John J. ("Black Jack") Pershing, leader of the American Expeditionary Forces. A striking dark bay horse with two white hind socks, Kidron captured the imagination of the American people because he was often used by Pershing in victory parades and seen in ceremonial photos. He became a symbol of all that was noble about the war, despite huge losses of equine and human alike. The news of Kidron's release from quarantine and his safe entry into the United States in 1920 made headlines across the country. On October 15, 1921, a plaque commissioned by the American Red Star was unveiled in the War Department in memory of the equine suffering during World War I. It reads: "This tablet commemorates the service and sufferings of the 243,135 mules and horses employed by the American Expeditionary Forces overseas during the Great World War, which terminated November 11, 1918, and which resulted in the death of 68,682 of those animals. What they suffered is beyond words to describe". A fitting tribute to their important services was given by the commander-in-chief of the American Expeditionary Forces, General John J. Pershing, who wrote "The army horses and mules proved of inestimable value in prosecuting the war to a successful conclusion. They were found in all the theaters of preparation and operation doing their silent but faithful work without the faculty for hoping for any reward or compensation." *(cited from* http://www.readex.com/blog/real-war-horses-america*)*

Quicksilver

Sir Percy Laurie's famous charger, Quicksilver, was a familiar sight to Londoners because of his many public appearances. Pictured *(right)* he wears his war medals on his bridle. Quicksilver took part in the Victory March and attended the funerals of all the great British WWI war leaders. His service started at the Somme in 1916 where he was wounded by shrapnel. He was, however, ridden continuously until the end of the war. After a spell in occupied Germany, he came to London in 1919 when his master joined the police. Quicksilver wears the Order of the Blue Cross and the 1916~18 Victory and General Service Medals. Quicksilver, when retired, went to the Ada Cole Memorial Stables in Hertfordshire. *(cited from* http://www.homefrontfriends.org.uk/wwihorse/wwihorsedespatch/page11.html*)*

The Unsung Heroes of World War I

Dolly and George Turner

George Turner joined the army in 1914 when he was underage at 16. With limited knowledge of horse care and management, he treasured the Blue Cross handbook and carried it with him throughout the war.

George survived some of the conflict's bloodiest battles, including the horrors of the Somme, but one act of bravery won him a Military Medal.

During an attack by the Germans, he was blown off his horse and, after pulling himself up, realized one of his other horses was badly injured by shrapnel. He took the ammunition from the injured animal, loaded some onto his horse (Dolly) and the rest on his own back and led both animals, under heavy fire, into a wooded copse in the middle of the field. He tied the injured horse up, determined to return while continuing on to the lines to deliver the ammunition. When he reached his comrades, George asked an officer to accompany him back to the corp to shoot the badly wounded horse, to put it out of its misery. The officer got halfway across the field and had to turn back due to the heavy shelling, but George pushed bravely on, reaching the copse and risking his own life to lead the injured horse back to the trenches so his officer could humanely end its suffering.

His granddaughter, Ruth Turner, said: "This act of courage shows how deep the bond was between my grandfather and his horses. Just three men survived the shelling that day but, despite great danger to himself, he refused to leave his horse to die a painful, lonely death. All this was watched by a French soldier through his binoculars, who reported it to my grandfather's commanding officer. As a result, he received the military medal for his bravery." *(cited from http://blogs.ancestry.com/uk/2011/11/07/guest-blogger-the-horrors-suffered-by-horses-during-wwi/)*

Jack Simpson and His Donkey

Jack Simpson Kirkpatrick was born in 1892 at South Shields in the north east of England. He came from a large family, being one of eight children. As a child during his summer holidays, he used to work as a donkey-lad on the sands of South Shields. He had a great affinity with animals, in particular, donkeys. Later he deserted ship in Australia when he heard of the war with Germany. Fearing that a deserter might not be accepted into the Australian Army, he dropped Kirkpatrick from his name and enlisted simply as John Simpson. He was to become Australia's most famous, and best-loved military hero.

In Perth on August 23, 1914, Jack was accepted and chosen as a field ambulance stretcher bearer. This job was only given to strong men, so it seems that his work as a stoker in the Merchant Marine had prepared him well for his exceptional place in history. He joined the 3rd Field Ambulance at Blackboy Hill camp, 35 km east of Perth on the same day.

On April 25, 1915, he, along with the rest of the Australian and New Zealand contingent landed at the wrong beach on a piece of wild, impossible and savage terrain now known as Anzac Cove. Attack and counter-attack began. During the morning hours of April 26th, along with his fellows, Jack was carrying casualties back to the beach over his shoulder–it was then that he saw the donkey. Jack knew what he had to do. From then on, he became a part of the scene at Gallipoli walking along next to his donkey, forever singing and whistling as he held on to his wounded passengers, seemingly completely fatalistic and scornful of the extreme danger. He led a charmed life from April 25, 1915, until he was hit by a machine gun bullet in his back on May 19, 1915. In these amazing 24 days, he was to rescue over 300 men down the notorious Monash Valley. His prodigious, heroic feat was accomplished under constant and ferocious attack from artillery, field guns, and sniper fire. *(cited from https://www.anzacs.net/Simpson.htm)*

The Unsung Heroes of World War I

San toy

It was said that San Toy (1890-1922) never missed a single day's duty through both the Boer campaign and the Great War. Working so well, and to the grand age of 28, his officers were keen to see him live out his retirement years in comfort. San Toy was one of the first 'War Horses' taken in by The Horse Trust's Home of Rest for Horses *(cited from http://www.horsetrust.org.uk/history/yesterday/war-horses/)*

Old Sam

Working horse Old Sam was a survivor of World War I but being just 11 years old in 1918 still had a lot of work left in him. He toiled for another 19 years, pulling firewood which his owner sold on the streets. Though his owner depended on and cared for him, life was hard, and there wasn't always enough money to look after Old Sam properly. When he could work no more, Mr. John Harris–the Magistrate of the Thames Police Court–asked The Horse Trust's Home of Rest for Horses to help. At last Old, Sam could enjoy some well-deserved rest and retirement at The Home of Rest. *(cited from http://www.horsetrust.org.uk/history/yesterday/war-horses/)*

Roger

Roger (1907- 1934) was a 15.2hh chestnut gelding thought to have been a German Officer's Charger. He was found wandering riderless on the battlefield during the infamous Battle of the Somme. A British Army Officer 'captured' Roger and got him to safety. Roger then served as that Officer's mount for the last 2 years of the war, getting him out of any number of 'tight spots' on the battlefield. After the war, the grateful officer bought Roger back to England and paid for him to enjoy a quiet and dignified retirement at The Home of Rest for Horses. *(cited from http://www.horsetrust.org.uk/history/yesterday/war-horses/)*

Jack

South Canterbury mounted riflemen - Major David Grant on his horse Jack (left). Major Grant was killed within 24 hours of arriving in Dardanelles. Jack continued on to France and was killed at the Somme.

The Unsung Heroes of World War I

The Sikh

The horse - named The Sikh - dodged shellfire and grenades as it delivered supplies to bloodied and battered troops during the First World War. She rode back and forth between the trenches and during Balkan battles with her devoted master Lieutenant A.C. Vicary of the Gloucestershire Regiment. Her extraordinary journey from Britain to the front line and her survival and trip home - in which she walked back from Russia - has now been unearthed by a war museum. Curators discovered she was a rare equine survivor of the Great War - and spent the rest of her life in Devon

The Sikh - who became a good luck omen with British troops - arrived at the front line with Vicary and the regiment's Second Battalion in Ypres. But when the war ended in 1918, she was all the way in Southern Russia - and had to walk all the way back to England. Chris Chatterton, the curator of the Soldiers of Gloucestershire Museum, unearthed the incredible story and is calling for a statue to honor the horse. He said: "I was reading a book about The Glosters and I came across a mention of The Sikh. I did some more digging, and it really is a remarkable story. She was viewed by many men in the Battalion as an omen of good luck." A statue to honor The Sikh would be great. We will certainly be doing something at the museum to commemorate her."

The brave horse was bred in Australia and sold to India. She arrived in North China with the 36th Sikh Regiment where the horse passed into the ownership of Lieutenant Vicary in 1913. When the battalion was given orders to return to Europe for war in November 1914, Vicary obtained special permission to take The Sikh with him. She was the only horse to accompany the Battalion from China, braving a treacherous eight-week boat journey from China to Europe. The Sikh spent the voyage in a makeshift open box on the deck - exposed to baking heat and typhoons as the ship traveled to the UK, dodging German battleships in the Mediterranean, according to the ship's log. While the soldiers were banned from sleeping on deck because of atrocious weather, the poor old horse was left out in the elements and only allowed to stretch her legs when officers went ashore at Hong Kong, Singapore, Port Said, and Gibraltar.

The Sikh was a loyal companion to Vicary throughout the entire First World War, supporting Vicary in the reserve lines and support trenches. The courageous pair led the 16th Gloucestershire Regiment in their victorious march through Serbia and Bulgaria. She managed to survive despite the desperate conditions for war horses. She proceeded with the Regiment to South Russia, before following them home through Turkey, Greece, Italy, and France once the war ended in 1918.

Vicary ended the war as a Lieutenant Colonel, having received a Military Cross and two Distinguished Service Orders medals for his gallantry. After her illustrious and adventurous life as a war horse, The Sikh died in peaceful retirement at Vicary's home in Devon. *(cited from https://www.express.co.uk/news/world-war-1/553386/Real-life-War-Horse-story-uncovered-in-animal-s-First-World-War-heroics)*

Dolly

Dolly was owned by Major-General Sir Andrew Russell *(left– not confirmed that horse is Dolly)* and on returning from the war was looked after by the Major's sister, Miss Gwendoline Russell, at the Russell family home of Heath House, Tuanui near Hastings. Dolly died in 1930 of natural causes. The Major, known as Guy to those close to him, was born 1868 in Napier to a farming family who had strong military ties. He spent his schooling years in the UK and then attended the renowned Sandhurst Military Academy. He helped raise the New Zealand Volunteer Force, commanded the Wellington Mounted Rifles Battalion and commanded the ANZACs. He died at home in 1960 aged 92 leaving his widow Gertrude and 4 children. All four of the New Zealand horses that made it home belonged to officers: Beauty to the late Captain Richard Riddiford, Bess to Captain Charles Powles, Dolly to Major-General Sir Andrew Russell, and Nigger to the late Lieutenant Colonel George King, who was killed at Passchendaele.

The initiative to bring home the four horses came from Major General Russell. Early in the demobilization process, he expressed a desire to bring home "a few" of the New Zealand Division horses, which had originally come from New Zealand, "owing to association over a long period of warfare." The list included Beauty, Bess and Dolly, which had left the country with the Main Body of the NZEF in 1914, and King's horse, Nigger, which had left in 1915 or 1916. The four horses were repatriated from France to England in March 1919 and subjected to 12 months' quarantine. They arrived back in New Zealand in July 1920.

Lieutenant Colonel King's daughter Nancy, who was 100 years old in 2014 remembered being at King's Wharf in Wellington to see the horses arrive off the passenger freighter, the SS Westmeath. The trip had been a traumatic one. Fire had broken out in the hold when the ship was six days out from Panama, destroying all the horses' feed and many of the soldiers' possessions. The soldiers fed their bread rations to the animals to keep them alive until they reached New Zealand. *(cited from http://birchhillmemorial.wixsite.com/ww1-ride/4-horses-returned and https://knowledgebank.org.nz/1703/1755/40403)*

Bess

Bess was among the several thousand riding horses that served with New Zealand's mounted forces in Sinai and Palestine. Bess was the only horse from the Middle East to return home to New Zealand. She was a four-year-old black thoroughbred. Bred by A. D. McMaster of Matawhero, near Martinborough, in 1910, she was by Sarazen out of Miss Jury. She was known as F. A. Deller's 'Zelma' prior to being presented to the New Zealand Army. 'Zelma' was allocated to the Wellington Mounted Rifles Regiment and selected by Captain Charles Guy Powles, who renamed her Bess.

Powles had previously served in the South African War. An officer in New Zealand's Staff Corps at the outbreak of the First World War, he became brigade major to the New Zealand Mounted Rifles Brigade (NZMR). He served with the NZMR in multiple campaigns during the war, becoming commanding officer of the Canterbury Mounted Rifles and ending the war as a lieutenant colonel.

Bess and Powles left New Zealand with the main body of the New Zealand Expeditionary Force in October 1914, bound for Egypt along with 3,815 horses. The horses were transported in rows of cramped stalls, some exposed to the weather. The men had to groom them, rub their legs to prevent swelling, and exercise them daily on coconut matting on the deck. About 3% of the horses died while being transported and their bodies were thrown overboard.

Very few horses were used at Gallipoli, so most, including Bess, remained in Egypt during 1915. In late 1915 the horses were reunited with their men as they returned from Gallipoli. Bess remained in the Middle East along with several thousand other New Zealand horses. They were assigned to the New Zealand Mounted Rifles Brigade which, as part of the ANZAC Mounted Division with Australian Light Horse brigades and Royal Horse Artillery batteries, served in the Sinai Campaign of 1916 and the Palestine campaign of 1917–18. Riding horses were used throughout by the mounted troops. The men were not cavalry, fighting from their horses. Rather, the horses allowed the riflemen to move rapidly to new positions. The men operated in groups of four and, when they dismounted to take part in actions like infantry, one man would be left behind to look after the horses.

The conditions on the ground in both the Sinai and Palestine were physically trying for the horses. They carried heavy loads–a fully loaded horse carried about 130 kilograms (286.6 pounds)–including rider, weapons, two bandoliers of ammunition, forage, blankets, food, and water. They had to travel long distances on difficult terrain, putting up with ticks, fleas and biting flies, shortages of food and water, and challenging weather–from extreme heat, burning sand and blinding dust, too cold nights and driving rain.

Inevitably horses lost condition. Some died, others were too weak to continue and were evacuated to hospital. In 1917, immediately before the Battle at Ayun Kara, some horses went without water for up to 72 hours. Horses also died as a result of wounds from enemy artillery fire or aerial attacks. The men became emotionally and physically dependent on their horses–and often used their shadows to get protection from the midday sun.

Shortly before the Palestine campaign ended Powles and Bess joined the New Zealand Division in France. After France, she served with Powles during the occupation of Germany's Rhineland. At the end of the war, an acute shortage of transport and quarantine restrictions related to animal diseases prevalent overseas prevented most horses from returning to New Zealand. Horses serving in Egypt were pooled with other British army horses. The fittest were initially kept, but most were not. Those fit for work were sold locally while those deemed unfit were killed. Some men tried to have their horses deemed unfit, rather than have them sold locally, concerned that the locals would mistreat them.

After the war, Bess was the model for the sculpture of a wounded New Zealand horse on a memorial to the Anzac mounted troops at Port Said in Egypt. Bess poses with trooper McKenzie *(photo p.82)* for the ANZAC Memorial sculptors after both had returned to New Zealand after the war. Trooper Clutha McKenzie had been blinded during action on Gallipoli and Bess was the only horse to return home that departed with the Main Body - both were selected to represent the New Zealand ANZAC's on the famed memorial.

Bess continued to serve Powles on her return to New Zealand in 1920 while he was a commander at Trentham and later as the headmaster at Flock House, an agricultural training school for dependents of war veterans. Bess produced several foals and died on land close to Flock House in 1934. Powles buried her at Flock House and erected a memorial. The square-shaped memorial, topped by a large rock, features two memorial plaques. One denotes the places where Bess served during and after the war. The other bears an Arabic inscription that translates as 'In the Name of the Highest God.' *(cited from https://nzhistory.govt.nz/media/video/bess-horse-great-war-story)*

The Unsung Heroes of World War I

Beauty

Captain Dick Riddiford was awarded the Military Cross during WWI for gallantry during an attack. Captain Riddiford, faced with the predicament of his beloved horse Beauty being disposed of after the war had ended, paid for Beauty to get back to England and a lengthy stint in quarantine. Tragically for Riddiford, he would not accompany Beauty home to New Zealand. While working as Aide-de-camp to General Officer Commanding, New Zealand Division (Major-General Sir Andrew Russell), he contracted the Spanish flu and died of pneumonia after a brief illness. He was buried at Brookwood Military Cemetery in Surrey. Beauty made it back to New Zealand and was met by Riddiford's sister and settled at "Westella" in Feilding, where the horse was ridden for 4 years until old age took its toll. *(cited from https://www.armymuseum.co.nz/museum-news/world-war-one-shared-histories/)*

Nigger

Nigger was an old black horse who traveled far with Lieutenant Colonel George King - Middle East, France, and Belgium - covering 6 years overseas, and returned to New Zealand with a shrapnel wound in one hind leg leaving a scar. Nigger was known for chewing his master's tunic at the left elbow whilst being held, and in fact, did this to the fur coat of his master's widow when she met Nigger at the dock on return to New Zealand. General Russell offered to look after Nigger, and with Dolly, both horses went to the Russell family home near Hastings where they lived until about 1930 when old age and rheumatism needed them being put to sleep. George Augustus King was born Christchurch 1885, educated at Christ's College, was a surveyor then worked on the family sheep farm in the Nelson area. In 1910 he married Annie; they had 2 children. He was always interested in the military and became a professional soldier in 1910. At Gallipoli, he was promoted to Major when commanding the Auckland Mounted Rifles Regiment. He was wounded in August 1915 at Hill 60 and awarded the Distinguished Service Order. He recovered and in February 1916 commanded the New Zealand Pioneer Battalion. Next, he served at the Western Front and at the Battle of Somme in the trenches. Under King's leadership, the Battalion's digging exploits gained a reputation for quality work and earned the nickname 'Diggers.'
In January 1918 a Bar was added to his Distinguished Service Order. In Belgium, 1917 during the Battle of Messines, Major King was awarded the Croix de Guerre and appointed Commander of the 1st Battalion Canterbury Infantry Regiment. He was the highest ranked New Zealander killed in the Battle of Passchendaele in 1917. He was 32 years old. A funeral was held at Ypres in Belgium. Sir Russell assisted his widow with the education of her 2 children Edward and Nancy *(cited from http://birchhillmemorial.wixsite.com/ww1-ride/4-horses-returned)*

Songster

Songster was the oldest horse to serve in France *(left with Trooper Bert Main)*

Fortunately, Songster escaped the horrific fate awaiting many of his four-legged and two-legged comrades and was even awarded medals for his exemplary service to his regiment. Days after the outbreak of war in August 1914, Songster mustered with troops in the Leicestershire Yeomanry in Loughborough's market square in preparation for battle. He was paired with Sergeant Main and shipped to France, but little is known about his time abroad.

Songster was known to have cantered to victory in a cross-country race, and he even broke free during one particularly heavy shell attack, but he returned to the frontline afterward.

While much of Songster's four years at war remains a mystery, what is clear is how highly his commanders thought of him as he was among only a handful of horses returned to the UK. Sergeant Main was so taken with his steed, he bought Songster and another former war horse from an auction in London and stabled him at West Beacon Farm in Woodhouse Eaves, Leicestershire.

Despite his advancing years, there was still plenty of pep left in the veteran steed. One day in 1920, he surprised a passing troop of the Leicestershire Yeomanry, who were marching to camp and passed Songster's field. At the sound of the bugle, he jumped over a 5-foot fence and approached them. They decided to take him to camp with them, and they did so every year until 1935. *(left: Songster with members of the Leicestershire Yeomanry at their annual camp in the mid-1930s).* Another of Songster's striking features was his age. He was thought to be about 30 years old when he died, but there are suspicions he may have been closer to 40, as it is not known exactly when he was born. In January 1940 Songster collapsed at West Beacon Farm. Songster was buried in the field where he grazed and his medals–two Mons stars, and General Service Medal, the Victory medal, and two Territorial Long Service medals and ribbons–were solemnly placed around his head. Sergeant Main handmade a wooden cross for his grave, painted in burgundy and with the words 'In Memorium Songster' emblazoned in gold, together with his regiment and his years in active service. *(cited from http://www.worcesternews.co.uk/news/9500861.Dad___s_war_horse_was_the_star_____100_years_ago/)*

The Unsung Heroes of World War I

Transport Horses

Corporal George Thompson, from Sunderland, was a transport driver with 7th Battalion Durham Light Infantry *(left: George–center back - with DLI Pioneers on the Marne, France July 1918)* during the Great War and used horses to deliver rations and supplies under enemy fire. As well as learning to drive heavy wagons, George was also able to ride bareback when required. In his memoirs he wrote; "Many a time I used to feel sorry for them," he said. "They used to stand out in all weathers, and sometimes up to their knees in mud. I can always say, while I had a pair of horses in France, I always did my duty to them. At the time we were stationed at St Jean our horses never had their harnesses off. We got shelled all day long while we were at St Jean. After a few days, we went further back, to a large field on the roadside near to Ypres. Here our horses had their harnesses taken off for the first time. We gave them a good cleaning down, and they looked a lot better. I remember one night when we [were] going up with rations they gave us in order to put our gas masks on, and we had to put them on our horses. We had some game on with them." *(cited from https://www.chroniclelive.co.uk/news/north-east-news/memoirs-world-war-one-soldier-10067710)*

Morning glory

Morning Glory was shipped to France from Brome County in Quebec's Eastern Townships in 1915. Her owner was Lt.-Col. George Harold Baker, known to friends and family as Harry. At the age of 38 he could easily have stayed home or worked behind the lines, but Baker volunteered to go overseas. When he went to France in 1915, he took Morning Glory with him. When Baker and his Mounted Rifles arrived in England, they were reclassified as infantry and sent to the trenches. The men were separated from their horses, which were sent to France. Morning Glory was lucky, avoiding the fate of so many of the other horses, such as dragging guns under fire through the mud. She caught the eye of a battalion commander who took her for his personal mount. Baker was separated from Morning Glory, but he saw his horse from time to time. He mentioned her in a letter home from Belgium dated May 5, 1916. "I saw Morning Glory day before yesterday; she is in the pink of condition. I hope someday to have her back." It was to be his last visit with her. Baker was killed around 8:30 p.m. on June 2, 1916, at Maple Copse in Sanctuary Wood during the battle of Ypres. The man who went to war thinking he would be leading the charge on his horse died instead in the mud in Flanders under unrelenting shellfire. Morning Glory came home to Canada in 1918 at the end of the war even though it was unusual for a horse to be shipped back from overseas. General Dennis Draper, a friend of Baker's, brought Morning Glory back to Quebec. "The horse never went into battle, which is why he came back to Canada," says Arlene Royea, managing director of the Brome County Historical Society, which operates a museum in Knowlton, Quebec. Morning Glory is buried behind Glenmere, the house at the family's summer home at Baker Pond, where a large bronze plaque is attached to a rock on a hill. The inscription is blackened in places and hard to read: "Here lies Morning Glory, a faithful charger who served overseas 1915-1918. Died 1936 aged 26 years." *(cited from http://www.cbc.ca/news/canada/morning-glory-canada-s-own-wwi-war-horse-1.1259736)*

The Unsung Heroes of World War I

Quotes and Stories

During the war, Harry Truman *(left)* was in command of Battery D, the "Dizzy D." They were wild, undisciplined, and proud of their reputation. They were a cohesive and athletic bunch, loyal to each other. Captain Truman was well liked and respected by his men; he was tough but fair. At Toul Hill in France in 1918, soldiers marched for ten days through deep mud, thick as paste. Gun wagons bogged down. Rain fell. Everyone was soaked, hungry, cold, and exhausted. Undernourished horses dropped dead in the stagnant muck. Worn out men reached out to steady themselves on the caissons, which was forbidden due to the extra burden it put on the already-overworked horses. A colonel noticed, and, in a fury, ordered the fatigued men to go faster and double-time up the hill. Rather than follow those orders, Truman ordered the battery off the road to rest for the remainder of the night. Somehow Truman avoided reprimand, and this incident and others demonstrated that he would protect his men and their horses. In return, he earned their loyalty. "We respected him," one soldier recalled, "because he earned it." *(cited from https://www.nps.gov/articles/harry-s-Truman-and-the-influences-of-his-service-in-world-war-i.htm)*

Lieutenant Henry Augustus "Harry" Butters, Jr., *(right)* was born in San Francisco, California, on April 28, 1892. He was educated both in his native country and in Great Britain. When World War I broke out, his heart was with the British, and he enlisted in the army. Harry stood six feet, two inches tall and impressed his comrades with his horsemanship. It is said that his exploits during the early months of the war were so outstanding that future British Prime Minister Winston Churchill, then serving as an officer on the Western Front, invited the young American to dinner in his fortified bunker. A few days before he became the first American citizen killed in World War I, Harry had spoken with a Roman Catholic chaplain and explained that he wished to have a Catholic funeral service and a quiet burial place marked by a cross with the simple phrase "An American Citizen" written on it.

Fritz Arnstein *(left: on the right with twin brother Rudi)* was a Jewish soldier in the Austria-Hungarian Army. In Marsha L. Rozenblit's broad study from 2001, "Reconstructing a National Identity, the Jews of Habsburg Austria during World War I" she writes that Jews' loyalty to Austria-Hungary stemmed largely from their chance to fight a "Jewish holy war" against czarist Russia, "home of pogroms and state-sponsored anti-Jewish oppression."

Fritz enlisted in the Habsburg cavalry as an "Einjahriger Freiwilliger," an officer's candidate and his roles at the front included dispatch riding, carrying messages among officers, crossing trenches and rivers under fire on horseback, and fighting on foot.: "We were allotted young and untrained horses, which had been requisitioned for war-service, and I received a young gelding with a highly independent character and rather sneaky. Once we were riding on an enormous potato field and practicing how to attack an imaginary enemy at the edge of the field, galloping at full speed, standing in our stirrups, body and head forward and the right arm with the sabre in the hand stretched over the head of the horse and shouting 'Hurrah!' with the full power of our young lungs to give us courage and scare the enemy. At the peak of the excitement, my horse suddenly stopped and bucked. I went right over his head and like a bullet into the potato field. Fortunately, I did not hurt myself and was back in my saddle before you could count to three. But the attack was stopped, the captain, a Count Schoenberg, turned around and when he saw my dirty face and uniform, he just remarked: 'Of course, the Jew.' He was later promoted to Oberleutnant — the equivalent to an American first lieutenant — in the fall of 1915. This was in Bukovina, the southwestern wing of the Eastern Front. Austria-Hungary's relatively liberal policies toward Jews enabled them to become officers in the military, unlike in Germany. *(cited from https://forward.com/news/205157/my-opas-story-of-world-war-ones-other-fight/)*

The most dangerous duty fell to Quartermaster Sergeants and teamsters who delivered supplies to the lines each night. They were confined to beaten down roads and anticipated by the enemy. The casualties among the men and mules who did this were heavy. There was no glory for those who did it. In the woods near a supply echelon of an infantry regiment there was an elaborately decorated grave with a large wooden cross with the following inscription: "Here lies poor Nelly of the Supply Company of the Sixteenth Infantry. Served in Texas, Mexico, and France. Wounded and killed near Villers Tournelles. She done her bit. Nelly was an army mule." *(cited from; "Where Have all the Horse Gone?" p 33)*

The Unsung Heroes of World War I

An A.D.V.S. (Assistant Director Veterinary Services) was giving instruction to a class of officers who were concerned with horses in the field, and one enterprising member of the class volunteered the information that he thought he knew all there was to know. He had, for instance, carefully read Horace Hayes' " Notes on Horse Management " and Fitzwygram's well-known book on " Horses and Stables." " Then," observed the A.D.V.S "I suppose you can tell me how many bones there are in a horse's foot." " There are three," promptly came the reply. The interrogator was naturally rather startled, and he had to investigate deeper and inquire the identity of the three. Our gallant officer obliged at once. " They are," he said, " ringbone, sidebone, and navicular "! He was not discharged the class that day. *(cited from: "The Horse and the War," p 80)*

It was believed that the mule had a sense of humor, but Captain Galtrey attributed the idea to the inquisitiveness of mules. "When not working (the mule) must be finding something to do with legs and mouth. I am reminded of an incident in an advanced mule line near Ypres. A number of friends were tethered in the open on a long rope, and a farrier was engaged in shoeing one. The mule this being attended to stood quietly enough, and the stooping farrier was performing his task so conscientiously well that he was naturally astonished when the next mule endeavored to take a mouthful from the seat of his breeches. Of course, he turned round sharply, as one would be stung in a particularly susceptible part of the anatomy, and, while his back was turned once more, the mule he had been shoeing gave him a sly kick on that same offending seat. Was that savagery? Of course not. Jack and Jenny were not vicious; they just wanted something to do." *(cited from The Horse and the War, pp.48-49)*

"The position over the rations for both men and horses was rather precarious. These were the days when we went without rations of any kind or water. The horses were more or less starved of water. On the retreat, we went to various streams with our buckets, but no sooner had we got the water halfway back to them, then we moved again. We had strong feelings towards our horses. We went into the fields and beat the corn and oats out of the ears and brought them back, but that didn't save them. As the days went on, the horse's belly got more up into the middle of its back, and the cry was frequently down the line, "Saddler - a plate and a punch!" This meant that the saddler had to come along and punch some more holes in the horse's leather girth to keep the saddles on." Gunner J W Palmer *(cited from http://www.firstworldwar.com/features/forgottenarmy.htm)*

There was a time in the early days of the war when the horse knowledge of such officers was more imaginary than real. For instance, an able and genial Assistant Director of the Veterinary Service, who was working in a particularly unhealthy part of the long line, told me a true story which amply illustrates with a saving grace of humor the square peg in the round hole. In the course of his visits a young infantry transport officer —such an officer may have about fifty animals in his care—complained of the poor quality of the oats. " What's the matter with the oats? " inquired the A.D.V.S. " Well, sir," was the reply, " they are so small; they get into the horses' teeth." " Ah, well, that's bad, very bad. Perhaps you'd better indent on ' Dados ' [a person who is known officially as the Deputy Assistant Director of Ordnance Supply] for some toothpicks "! Of course, the zealous transport officer meant well. But the best part of the story is that a day or two later the boy was ordered to replace a casualty in the Une, and the first time he went over the top he won the Military Cross. Clearly, it was a case of a square peg has been in the round hole. *(cited from: "The Horse and the War," pp 77-78)*

Our ammunition wagon had only been there a second or two when a shell killed the horse under the driver. We went over to him and tried to unharness the horse and cut the traces away. He just kneeled and watched this horse. A brigadier then came along, a brass hat, and tapped this boy on the shoulder and said, "Never mind, sonny!" The driver looked up at him for a second and all of a sudden he said, "Bloody Germans!" Then he pointed his finger, and he stood like a stone as though he was transfixed. The Brass Hat said to his captain, "All right, take the boy down the line and see that he has two or three days rest." Then he turned to our captain and said, "If everyone was like that who loved animals we would be all right." ***Gunner H Doggett*** *(cited from http://www.firstworldwar.com/features/forgottenarmy.htm)*

The intrepid mule – as bombs blast nearby and gunfire is all around, the mule does his job.

Musings of a Mule

I am only a common or garden mule
Who was bred in the U.S.A. I was born in a barn on a Western farm
Many thousands of miles away
From where I am munching a Government lunch
At Great Britain's expense today.

With dozens of others, I knew, and have seen,
In my Little Grey Home in the West,
Where the grazing was succulent, luscious and green,
And Life was a bit of a jest,
I have sniffed the salt breeze blowing over the seas
And I've landed in France with the rest.

The journey was horrid—a horrible dream
Was the loading — its shindy and row
And the people expecting a moke to be keen
To swarm up a frightening " brow "
And slither down ramps that were greasy and damp
To a standing unfit for a cow.

They packed us like herrings 'way down in the hold,
With never thought nor a care
For animals worthy more Government gold

Then all of the rest who were there;
And the best spot, of course, was reserved for the horse,
Who had to have plenty of air.

Well, we jibbed, and we strafed, and we kicked the Light Draught
And I planted my heels in the hide
Of a man on the ship who was flicking a whip
And whose manners I could not abide;
But I've traveled so often since then in the trucks
I have learned how to swallow my pride,
And I go where I'm put without lifting a foot
For a rag song and a dance on the side.

Many months at a time I was up at the Somme
In the rain and the mud and the mire:
We were " packing " the shells to the various Hells
In the dips of the vast undulations and dells
Where the field guns were belching their fire.

It was very poor sport when the forage ran short
First to eight and then six pounds a day.
But we managed to live on the blankets they brought,
Though blankets I now think, and always have thought.

The Unsung Heroes of World War I

Are but poor substitution for hay.

I remember a week when we played hide and seek
With the shrapnel the Boches sent over:
I remember the night when they pitied my plight,
And pipped me, and put me clean out of the fight
With a " Blighty "—then I was in clover.

For they dressed me and sent me quick out of the line
To a hospital down at the Base,
Where the standings were good, and the weather was fine
And the rations were not a disgrace:
There, just within the sound of the Heavies, I found
La France can be quite a good place.

And now I've recovered —I'm weary and thin. '
And I'm out of condition and stale,
My ribs and my hips are too big for my skin
And I've left all the hair on my tail

On the middlemost bar of the paddock, I'm in.
For they turned me out loose, as I'm frail.

Now the life in a paddock according to men
Is a sort of a beautiful song
Where animals wander around and can squander
The time as they wander along,
With nothing to worry them, nothing to do
Except for food intervals daily; but you
Can take it from me they are wrong,
For paddocks are placed conducive to thoughts
That settle unbid on the brain,
And often I find them to follow a kind
Of a minor-key tune or refrain As I doze for an hour in the afternoon sun
Or I stand with my rump to the rain I dream of the barn
on my Illinois farm
And I long to be back there again.
(cited from: "The Horse and the War", pp 51-52)

Graves of two men who were among the last of the 1st Light Horse Regiment killed in the Middle East campaign. Even in death, the regiment regarded the soldier and his horse as a team. (cited from http://www.diggerhistory.info/pages-conflicts-periods/ww1/lt-horse/first-must-horse.htm)

The caption written on the back of this photo by the soldier is: "She may be stupid but I love her very much." (cited from: https://www.thedonkeysanctuary.org.uk

The Unsung Heroes of World War I

The Homefront Without Horses and Mules

*c. 1915 An elephant from a Belgian zoo is put to work on a Belgian farm during World War I.
(cited from https://mashable.com/2015/07/29/elephant-farming-war/#49M0k0_rMmq2)*

When World War I erupted in Europe, Britain did not just draft men to fight. The government conscripted horses and mules to aid the war effort. This left a shortage of beasts of burden to keep up with farm work on the home front, meaning farmers had to resort to more unconventional help. Circus elephants, *(left: Lizzie pulling a cart up and down the cobbled streets of Sheffield)* with the strength of several workhorses, were recruited to help plow fields, stack hay, and cart munitions and other supplies around cities. *(cited from https://mashable.com/2015/07/29/elephant-farming-war/#49M0k0_rMmq2)*

The huge growth of munitions output had ripple effects across the war economies: not least on agriculture and food supply. In the Central Powers, the blockade exacerbated shortages of fertilizer, tools, and lighting fuel. Allied farmers suffered less in these respects, but they too were deprived of draught animals and especially of labor. In most countries, armies recruited disproportionately from the countryside. Many wartime family farms were run by the wives, assisted by their children and sometimes by migrant workers and prisoners of war. *(cited from https://www.bl.uk/world-war-one/articles/the-war-effort-at-home)*

During World War One, 23,000 British women were recruited to work full-time on the land, to help replace men who had left to fight in the war. *(left: Women's Land Army recruitment poster by the artist J Walter West, emphasizes the work done by the Forage Department of the WLA. Source: Imperial War Museum.)* This form of National Service for young female civilian farm workers was called the Women's Land Army. There were three sections of the Women's Land Army: 1Agriculture 2) Forage (haymaking for food for horses) 3) Timber Cutting. The United States was quick to follow with a Woman's Land Army *(cited from http://www.womenslandarmy.co.uk/overview/)*

The Unsung Heroes of World War I

Many men from farms went off to war. Adding to the labor shortage on farms was a reduction in the number of horses for use on the land, and a shortfall in the number of tractors suitable for farming. Under the Defense of the Realm Acts, the Board of Agriculture had the power to requisition horses, machinery, and supplies in order to improve the food situation. Large numbers of horses were required for the army, leaving insufficient animals. Farms of less than 250 acres had a small workforce. Usually, only a few men and horses, and consequently shortages were felt more immediately than on larger farms. In 1916 the Devon War Agricultural Committee reported substantial shortages in horsepower to the Board of Agriculture. The complaint was based on earlier reports from farmers that when horses did become available, they were often of substandard quality for farm work. Many were old or sick, or they were light vanners that were not of the proper weight and size to pull a plow. Ploughman F. Goldman explained that 'the poor horses are now being overworked to an awful extent. They are laden and fatigued almost to death. This is a terrible way to farm'. Cecil Doidge, an owner-occupier from Holsworthy, explained that 'We cannot be expected to increase production with limited manpower, short of horses, and plows that have proven useless. Besides, what should we use to pull the plows? What should we do with tractors that are unsuited for the land in Devon?' At the end of 1916, an inadequate supply of horse harnesses and a deficiency of skilled plowmen made the problems in Devon even acuter. It was not until June 1917 that the Board of Agriculture took steps to protect the supply of horses for agricultural use when, under Regulation 2T of the Defense of the Realm Act (DORA), the sale of farm horses was prohibited without a license from the local agricultural board. With horses in short supply, farmers had to rely more on machinery. Until 1917, however, farm machinery was in short supply, often shared between several farms, and primarily confined to larger farms.
(cited from http://www.bahs.org.uk/AGHR/ARTICLES/58_1_5_White.pdf)

The Belgian economy couldn't operate without horses. In the transportation sector, horses still pulled a number of trams, carts in the cities and villages and even the stagecoaches that were put back into local service in order to overcome the fact that it was impossible for civilians to travel by train. Horses were indispensable for agriculture, both for plowing and for the transportation of foodstuffs. No plant or manufacturing facility could operate without them, both for bringing in the raw materials and for carrying away the finished products. The horses pulled the trucks at the bottom of the mines, brought the ore to the surface and to the loading docks, and pulled the barges on the canals. Over and above the pre-war needs, horses were also indispensable for the CNSA (National Relief and Food Committee) in order to supply the distribution points with the foodstuffs needed for the survival of the population. In Brussels and its suburbs, it wasn't unusual to see men harnessed to heavily laden horse carts. *(cited from https://www.rtbf.be/ww1/topics/detail_the-horse-an-essential-participant-of-the-great-war?id=8358614)*

Armistice and the Post-War World

On the 11th hour

of the 11th day

of the 11th month

we will remember them.

For Horses and Heroes, the Poppies Grow" by Jacqueline Hurley

The Unsung Heroes of World War I

What Ended the war?

At the start of 1918, Germany was in a strong position and expected to win the war. Russia had already left the year before which made Germany even stronger. Germany launched the 'Michael Offensive' in March 1918, where they pushed Britain far back across the old Somme battlefield. However, their plan for a quick victory failed when Britain and France counter-attacked.

By 1918, German citizens were striking and demonstrating against the war. The British navy blocked German ports, which meant that thousands of Germans were starving and the economy was collapsing. Then the German navy suffered a major mutiny. After German Emperor Kaiser Wilhelm II abdicated on November 9, 1918, the leaders of both sides met at Compiegne, France. The peace armistice was signed on November 11th. in the Forest of Compiègne, in French Marshal and Allied Supreme Command Ferdinand Jean Marie Foch's private railcar, Western Allied and German commanders and politicians signed the agreement to cease hostilities. *(left)* The French turned the car into a monument. When Hitler returned to France with Nazi Germany's invasion a little over two decades later, he forced the French to sign the Armistice in the exact same car. *(right)* He later ordered it blown up to avoid a similar symbolic gesture returned in kind to Germany.

By the time it was said and done, four empires — the Russian, the Ottoman, the German, and the Austro-Hungarian had collapsed because of the war. In 1919, The Treaty of Versailles officially ended the War. But the Treaty was brutal towards Germany. Much of the Armistice was built on Woodrow Wilson's "14 Points" — requiring that Germany accepts full responsibility for causing the war; make reparations to some Allied countries; surrender some of its territories to surrounding countries; surrender its African colonies and limit the size of its military. Though the Armistice ending World War I was essentially a German surrender, it still can't quite be called true peace, and even if one calls it peace, it wasn't truly a full defeat, a complete victory. General John J. Pershing, commander of the American Expeditionary Force, for one example, wanted to crush the Germans, believing that only their total defeat would keep the nations of Europe from fighting again before long.

The Treaty also established the League of Nations to prevent future wars. The League of Nations helped Europe rebuild, and 53 nations joined by 1923. But the U.S. Senate refused to let the United States join the League of Nations, and as a result, President Wilson (who had established the League) suffered a nervous collapse and spent the rest of his term as an invalid. Although Germany joined the League in 1926, continuing resentment because of The Versailles Treaty caused them to withdraw (along with ally Japan) in 1933. Italy withdrew three years later. The organization subsequently proved helpless to stop German, Italian, and Japanese expansionism.

Some might argue that World War I never had an effective ending, but the battles just stopped. World War II may never have happened if not for World War I, because had the Germans not been beaten down so badly by the demands of The Treaty of Versailles, Adolf Hitler may not have risen to power in the 1930s and convinced the Germans to fight to regain their dignity and place in the world.

At last. "The War to End All Wars" was over. In the cites people celebrated, but on the battlefields, for the most part, the troops were somber, quiet and exhausted after a long war. Jörn Leonhard, a British corporal, recalled "… the Germans came from their trenches, bowed to us and then went away. That was it. There was nothing with which we could celebrate, except cookies."

(cited from http://www.bbc.co.uk/schools/0/ww1/25403869 and https://www.cliffsnotes.com/cliffsnotes/subjects/history/how-did-world-war-i-start-and-end and https://www.warhistoryonline.com/world-war-i/ed-ok-10-facts-end-WW1-armistice-November-11th.html and http://ww2today.com/the-french-sign-the-armistice)

The Unsung Heroes of World War I

Bringing the Boys and the Horses Home

Britain had around 900,000 horses overseas at the time of the armistice and, while the repatriation of soldiers proceeded with as much haste and efficiency as could be made possible, the return of horses was viewed as a needless encumbrance to this priority. Far fewer horses were going to be required for Britain's peacetime army, and it was deemed more productive to leave most of them where they were. The reward for the majority of these animals was therefore to either be auctioned off and to spend the remainder of their days toiling in a foreign field or, for those animals in less than good condition, to be slaughtered, with their meat being used both to boost domestic food stocks and to help feed the tens of thousands of POW's whose welfare was the responsibility of the victorious nations. Horses and mules were auctioned off to farmers on the continent for an average of 37 pounds (= to 1,796 pounds in 2018). The oldest and most worn out horses were sent to the knacker's yard for meat and fetched 19 pounds (= to 922 pounds in 2018) - a necessary move when severe food shortages hit Europe at the end of the war. Thousands of Australian horses did not return home and were used by the British Army in India. General Jack Seely (owner of Warrior) was so incensed at the treatment being given to those horses that remained overseas that he wrote to Winston Churchill demanding that something is done regarding the intolerable delay in their transportation home

War Office documents found in the National Archives at Kew show that tens of thousands of the animals were at risk of disease, hunger and even death at the hands of French and Belgian butchers because bungling officials couldn't get them home when hostilities drew to a close. Churchill, then

aged 44 and Secretary of State for War was an avid horse lover *(left photo from 1950)* reacted with fury when he was informed of their treatment and took a personal interest in their plight after the war. He secured their speedy return after firing off angry memos to officials within his own department and at the Ministry of Shipping. In a strongly worded missive dated February 13, 1919, Churchill told Lieutenant-General Sir Travers Clarke, then Quartermaster-General: 'If it is so serious, what have you been doing about it? The letter of the Commander-In-Chief discloses a complete failure on the part of the Ministry of Shipping to meet its obligations and scores of thousands of horses will be left in France under extremely disadvantageous conditions.' Churchill's intervention led to extra vessels being used for repatriation, and the number of horses being returned rose to 9,000 a week. It is mainly through the direct efforts of Churchill that around 60,000 horses were eventually returned to Britain, still a pitifully small number of the total that had survived the war only to be discarded by the army in the most heartless way conceivable. The Army Veterinary Corp must be excluded from such criticism, having provided sterling service to the two and a half million animals it treated during the war, with 80% of all injured animals cared for by the corps being able to return to duty. In recognition of the veterinary corps service throughout the war, the "Royal" prefix was granted to them on November 27th, 1918. *(cited from http://nerdalicious.com.au/history/war-horses-britains-equine-army-of-the-first-world-war/ and http://www.dailymail.co.uk/news/article-2080777/Churchills-mission-rescue-war-horses-officials-bring-tens-thousands-home.htm and http://bam.files.bbci.co.uk/bam/live/content/zqn9xnb/transcript)*

There were 122,000 horses exported from Australia during WWI. Australians were very proud of their War Horses *(left: war horse parade-Brisbane 1918)*. At the end of 1918, 11,037 were still in use with the Australian Light Horse units in the Middle East. It was for quarantine reasons that wartime animals could not be brought home. These were classified by age and condition into four groups. Those not fit for further use (2,853) were destroyed. Their manes and tails were shorn as horse hair could be sold and their shoes removed. The remainder were sold to the British Army, the Indian Army, and the Finnish Army. 600 mares were sent to England for breeding purposes. One of the most common myths about the Australian Light Horse is that they [servicemen or soldiers] shot all of their horses at the end of WWI. Only 250 were allegedly destroyed by the rider without permission for they feared the horses would be mistreated if sold locally. The situation was slightly more promising for the New Zealand Division's horses in France. They were similarly pooled with other British army horses and then killed, sold or retained. But the odds were far better: around 100,000 of the British army's nearly 400,000 horses in France were eventually repatriated to England. *(cited from http://blogs.slq.qld.gov.au/ww1/2014/04/29/horses-at-the-end-of-wwi/ and https://commons.wikimedia.org/wiki/File:StateLibQld_1_152275_War_Horse_Day_procession_in_Brisbane,_1918.jpg and https://nzhistory.govt.nz/war/nz-*

War Memorial in Hyde Park, London, dedicated to animals who have died in the war. The inscription says 'They had no choice.'

The Unsung Heroes of World War I

American horses also did not come home. There was apprehension in the United States that animals serving with the A.E.F. might be returned to the U.S. and carry some disease or infection. W.H. Butler of the Ohio Percheron Breeders' Association wrote a letter to the War Department expressing this concern. On January 30, 1919, the War Department directed that: "No public animals belonging to the military forces will be imported from Europe to the United States." It was directed however that up to 200 private mounts could be imported subject to 90 days in quarantine in Europe, shipment in isolation, and a further 90 days quarantine in the U.S. The officer was also required to verify that he was the owner of the horse. Some of the U.S. horses that remained were transferred to the Remount Service. Others were sold at 600 public auctions conducted by the A.E.F. in France. The French Government purchased 33,045 animals for distribution to inhabitants of devasted regions. The Polish government bought 5,000 and Belgium bought 400 cavalry horses. Other purchases came from England, Serbia, Switzerland, and Czechoslovakia. The Third Army assigned to Luxembourg and occupation duty in German Rhineland was provided 50,340 animals. In 1921 there was erected in the State, War and Navy Department Building in Washington a memorial tablet to commemorate the services of American horses and mules in the war.   *(cited from: "Where Have All the Horse Gone," pp 54-58)*

After the First World War the role of the horse changed forever due to the increased mechanization of the modern world and, despite its future deployment in areas of rugged or inhospitable terrain, it would never again play such a significant part in military conflicts. Science has given man easier and more direct means to kill one another without involving the horse, and for that, at least, we should be grateful. This noble beast will never again have to suffer the horrors of modern warfare only to be slaughtered upon the expiry of its usefulness by the very people it had trusted and had given its all for. *(cited from http://nerdalicious.com.au/history/war-horses-britains-equine-army-of-the-first-world-war/)*

This touching shot shows some 650 soldiers standing in a formation, which from a bird's-eye view perfectly resembles a horse's head, neck, and a noseband. The moving shot is a striking tribute to the brave horses who died during the First World War. It is believed the picture was taken by officers of the Auxiliary Remount Dept No.326 in Camp Cody, New Mexico in 1915.(cited from https://www.express.co.uk/news/history/615101/World-War-One-horses-killed-Remembrance-Day-November-11)

The Unsung Heroes of World War I

Dorothy Brooke

The letter below was written by Dorothy Brooke and published in The Morning Post (now the Telegraph) in 1931, alerting the British public to the plight of ex-war horses.

"There have been several references lately in the columns of The Morning Post as to the possibility of raising a memorial to horses killed in the War. May I make a suggestion?

Out here, in Egypt, there are still many hundreds of old Army Horses sold of necessity at the cessation of the War. They are all over twenty years of age by now, and to say that the majority of them have fallen on hard times is to express it very mildly. Those sold at the end of the war have sunk to a very low rate of value indeed: they are past 'good work' and the majority of them drag out wretched days of toil in the ownership of masters too poor to feed them–too inured to hardship themselves to appreciate, in the faintest degree, the sufferings of animals in their hands.

These old horses were, many of them, born and bred in the green fields of England–how many years since they have seen a field, heard a stream of water, or a kind word in English? Many are blind–all are skeletons.

A fund is being raised to buy up these old horses. As most of them are the sole means of a precarious livelihood to their owners, adequate compensation must, of necessity, be given in each case. An animal out here, who would be considered far too old and decrepit to be worked in England, will have before him several years of ceaseless toil–and there are no Sundays or days of rest in this country. Many have been condemned and destroyed by the Society for the Prevention of Cruelty to Animals (not a branch of the RSPCA), but a want of funds necessitates that all not totally unfit for work should be restored to their owners after treatment.

If those who truly love horses–who realize what it can mean to be very old, very hungry and thirsty, and very tired, in a country where hard, ceaseless work has to be done in great heat–will send contributions to help in giving a merciful end to our poor old war heroes, we shall be extremely grateful; and we venture to think that, in many ways, this may be as fitting (though unspectacular) part of a War Memorial as any other that could be devised."

Signed–Dorothy E. Brooke–1931

Dorothy had loved horses since childhood and was an accomplished horsewoman. The defining moment in Dorothy's life came when she arrived in Cairo, the newly married wife of British cavalry officer Brigadier Geoffrey Brooke, in October 1930. From this moment her life became dedicated to the welfare of working horses and donkeys and, despite many obstacles and objections, her determination to make a difference ensured that the original Old War Horse Memorial Hospital was founded in 1934.

On arrival in Egypt, Dorothy Brooke was determined to find the surviving ex-warhorses of the British, Australian and American forces. These brave and noble horses were sold into a life of hard labor in Cairo when the conflict ended. Searching for them throughout Cairo, Dorothy was appalled to find hundreds of emaciated and worn-out animals desperately in need of help. She wrote the above letter to the Morning Post exposing their plight. The public was so moved they sent her the equivalent of £20,000 ($27,917) in today's money to help end the suffering of these once proud horses. Within three years, Dorothy Brooke had purchased five thousand ex-warhorses. Most were old, exhausted and had to be humanely put down. But thanks to her compassion, they ended their lives peacefully.

Dorothy Brooke knew thousands of hard-working horses, donkeys and mules still suffered so in 1934 she founded the Old War Horse Memorial Hospital in Cairo, with the promise of free veterinary care for all the city's working horses and donkeys. The Brooke Hospital for Animals was born. Dorothy Brooke wrote diaries throughout her time in Egypt. They chart Dorothy's dedication and determination to set up the hospital. Dorothy continued to work for her charity until her death on June 10, 1955. She was buried in her adopted home of Cairo, but the strong family association that founded the Brooke still continues today. Family members are still involved with the charity, ensuring that the spirit of Dorothy Brooke lives on. The Brooke aids over two million working horses, donkeys and mules across Africa, Asia, Latin America and the Middle East. Their staff includes vets, animal welfare experts, and advocacy and development specialists. Learn more about The Brooke at www.thebrooke.org *(cited from https://www.thebrooke.org/)*

The Unsung Heroes of World War I

Devastation

The devastation of World War I spearheaded a search for a newly regulated urban order. The war devastated European society and hastened new ideas of city building which would come to dominate urban planning and policies for urban renewal and reconstruction in the twentieth century. World War I reorganized the prevailing notions of city life in the United States and Europe. The new economy of war and the necessity of housing industrial workers in the United States and rehousing and reconstituting devastated cities and populations in Northern Europe changed urban life. Paris sustained the first strategic air attack of the war in August 1914. In addition to the psychological impact of the German campaign against Paris, the city began to take on a "new geography" as a result of the war. Mechanization was seen as a positive vehicle for urban modernization and advanced, civilized society. Mass production, construction technology and the essential building materials of structural steel, glass, and concrete evolved, and life in cities changed; it demanded a complete reorganization of beliefs and a turning away from the past; the use and need of the horse in the city. *(cited from http://net.lib.byu.edu/~rdh7/wwi/comment/slavitt.htm)*

The Changing World–Goodbye to the Horse

As Europe changed after the War, so did North America.

North Americans employed 4 million horses in 1840 for agricultural work and travel. By 1900 they were harnessing more than 24 million (a six-fold increase) to plow fields, as well as pull street trolleys, drays, brewery wagons, city vehicles, omnibuses, and carriages. For every 3 people there trod 1 working horse in the U.S. There are now 1.3 people for every car in the U.S.

By 1890 New Yorkers took an average of 297 horse-car rides per person a year. Today, they hail an average of 100 cab rides.

In a New York City traffic study undertaken in 1907, horse-drawn vehicles moved at an average speed of 11.5 mph. (left: New York prior to WWI). A similar study conducted almost 60 years later found that automobiles moved through the city's business district at an average speed of only 8.5 mph. Then along came the combustion engine. But it took the automobile and tractor nearly 50 years to dislodge the horse from farms, public transport, and wagon delivery systems throughout North America. *(cited from https://thetyee.ca/News/2013/03/06/Horse-Dung-Big-Shift/ and https://parkcityhistory.org/wp-content/uploads/2012/04/Teacher-Background-Information.pdf)*

Many of the new technologies developed during World War I eliminated the need for horses. The internal combustion engine took from the horse the distinction it had enjoyed over the centuries as the sole, or primary, source of mobile energy and swept the horses from the road, the farm and the battlefield. Not only were horse's jobs gone but gone too were the jobs of all those involved with the horse economy; the Teamsters, the hostlers, the grooms, the farriers, wheelwrights, carriage painters, carriage builders, draymen, liverymen, makers of saddle, whips, blankets and other horse clothing, manure transporters. Gone too were the jobs of the farmers in the belt of farms around each city, growing the forage to feed the city horses, the hay and grain dealers. The "job" of the horse became one of companion and teammate in athletic endeavors; hopefully to never again have to endure the horrors of war on such a large scale. *(cited from: "Where Have All the Horse Gone?", pp 185-86, back cover)*

The Unsung Heroes of World War I

Horses Are Still Our Heroes

Special Operations Soldiers and Airmen join bestselling author Doug Stanton at the America's Response Monument, following the rededication ceremony at Liberty Park in New York City in September 2016. Stanton, the author of The Horse Soldiers, attended the rededication ceremony along with more than 500 attendees. The statue sits in overwatch of the 9/11 Memorial. Pictured from left to right are Chief Warrant Officer 2 Brad Fowers, Master Sgt. Keith Gamble, Maj. Mark Nutsch, Air Force Lt. Col. Allison Black and Stanton. (U.S. Army photo by Cheryle Rivas, USASOC Public Affairs.) (Photo Credit: Ms. Cheryle Rivas (USASOC))

Modern-Day Horse Soldiers

Secretary of Defense Donald Rumsfeld revealed this photo to the public on November 16, 2001. It shows U.S. soldiers with General Dostum and other Afghan fighters riding on horseback in the Darya Suf Valley (arrows added to specify which are U.S. Soldiers).

"Horse Soldiers," is a book written by Doug Stanton about a band of Special Forces soldiers in Afghanistan who rode horseback in the war against the Taliban after 9/11. Jerry Bruckheimer produced the movie, "12 Strong", based on the book.

In response to the September 11, 2001 attacks, the elite U.S. Special Forces unit, Operational Detachment-Alpha 595 (ODA 595 for short), was one of three teams of Special Forces soldiers sent into Afghanistan. At first, their mission was one of personnel recovery, tasked with rescuing any pilots shot down during the air war in Afghanistan. However, the mission quickly changed and became about convincing ethnic leaders (some were more similar to warlords) to join forces with them to fight their common enemy: the Taliban and its Al Qaeda allies.

The Unsung Heroes of World War I

The CIA provided intel on which ethnic leaders to work with, including Afghan General Rashid Dostum. Once the ODA 595 linked up with Dostum, they were to "render the area unsafe for the Taliban and terrorist activity," says Green Beret Mark Nutsch. ODA 595 was an experienced, mature team of Green Berets that had recently worked with special operations forces in Uzbekistan, Afghanistan's northern neighbor. The team had been working together for two years, and the average age was 32 years old. Each member had an average of eight years experience and most had combat experience in either Desert Storm, Kosovo or Somalia. As for team leader Mark Nutsch, he had no actual combat experience prior to the mission. Originally, the Special Forces team, ODA 595, was supposed to be in Afghanistan on September 14, just three days after the attacks. However, at least 3 false starts delayed their deployment for almost a month; they finally left Kentucky on October 5, 2001.

The soldiers on the U.S. Special Forces teams were used to state-of-the-art warfare. However, the rugged Afghan terrain forced them to adopt the rudimentary practices of the Afghan horse soldiers. This included using horses to navigate the mountainous region. Officials had been unaware that Dostum's army still used a horse cavalry. It was indeed as much of a surprise to the team. Mark Nutsch, had worked on a cattle ranch and competed in collegiate rodeos when he was younger and was one of the few members of the 12-man team who had experience riding horses.

"They didn't expect us to survive," Mark Nutsch said. "The threat of capture, torture was very real." Furthermore, they were grossly outnumbered. There were roughly 200 paid Afghan soldiers under General Dostum's command and an undetermined number of part-time militia. They were potentially facing about 50,000 Al-Qaeda and Taliban fighters. Also, there was hardly any assurance that the team would be safe with the Afghan fighters they were working alongside. And if they became overwhelmed by the enemy, little could be done to save them since they would be roughly nine hours away from help. "For all of our teams, the risk was extraordinary," Lt. Gen. John F. Mulholland said. "If they got in trouble, there was very little I could do and nothing I could do quickly. We accepted a huge amount of risk." All the members ODA 595 survived!

It took just a few months for American and Afghan forces to defeat the Taliban in Afghanistan; thus, ousting them from power. The actions of the ODA 595 helped lay the groundwork for that victory. The "Horse Soldiers" were able to liberate Afghanistan's fourth largest city, Mazar-i-Sharif, in just three weeks. Mazar-i-Sharif was the stronghold of the Taliban's northern force and once liberated, the northern provinces quickly fell. From there, American and Afghan forces would liberate Kabul in the east, Herat in the west, Kandahar in the south, and Jalalabad, resulting in the Afghan forces taking control of the country from the Taliban. In all, less than 100 Special Forces soldiers toppled the Taliban government. Military historians have called it one of the most successful unconventional warfare campaigns in U.S. history. *(cited from http://www.historyvshollywood.com/reelfaces/12-strong/)*

Wounded Warriors

An equestrian therapeutic pilot program was started at Fort Meyer, Virginia by volunteers from Caisson Platoon Old Guards. Wounded soldiers were trained as horse leaders and side walkers with the aim of improving their balance and coordination and reaping other benefits of physical therapy through horse riding.

From May 12 to June 2, 2006, Caisson Platoon helped wounded and amputee soldiers from Iraq and Afghanistan in what is known as "soldiers for soldiers" model. Not only were the wounded soldiers taught, but the horses were carefully handpicked and trained as well.

During the initial session, the rider's skills were evaluated, and a baseline was established. The next sessions progressively challenged the riders as they were given tasks with varying levels of difficulties. Tasks ranged from working with horses, relay racing and barrel trot racing where the riders were pushed towards improving their skills in every session.

Working alongside with the volunteers was a Walter Reed Medical Center occupational therapist to assess every rider's skill in certain areas before and after every ride. From what the therapist observed, he or she noted that when the riders adjusted to the horse's motion, it helped with core strengthening the lower back and hips. At the same time, it took pressure and pain off the amputated limbs. A new core for balance and a revived sense of control were developed by the riders.

The equestrian therapeutic program is now popularly known as the Horses4Heroes program. Different therapeutic riding programs for the wounded soldiers can be found all over the USA. The program is being undertaken under the Professional Association of Therapeutic Horsemanship International (PATH International) after the successful pilot program in Fort Hood and Fort Meyer. *(cited from https://www.equestriantherapy.com/equine-therapy-wounded-soldiers/)*

Facts and Figures

TOTAL MOBILIZED AND CASUALTIES

Country	Total Mob. Forces	Wounded	Killed	Prisoners and Missing	Total Causalities	Casualties as % of Forces
Allied and Associated Powers						
Russia	12,000,000	4,950,000	1,700,000	2,500,000	9,150,000	76.3
British Empire	8,904,467	2,090,212	908,371	191,652	3,190,235	35.8
France	8,410,000	4,266,000	1,357,800	537,000	6,160,800	73.3
Italy	5,615,000	947,000	650,000	600,000	2,197,000	39.1
US	4,355,000	204,002	116,516	4,500	323,018	7.1
Japan	800,000	907	300	3	1,210	0.2
Romania	750,000	120,000	335,706	80,000	535,706	71.4
Serbia	707,343	133,148	45,000	152,958	331,106	46.8
Belgium	267,000	44,686	13,716	34,659	93,061	34.9
Greece	230,000	21,000	5,000	1,000	27,000	11.7
Portugal	100,000	13,751	7,222	12,318	33,291	33.3
Montenegro	50,000	10,000	3,000	7,000	20,000	40.0
TOTAL	42,188,810	12,800,706	5,142,631	4,121,090	22,062,427	52.3
Allied and Associated Powers						
Germany	11,000,000	4,216,058	1,773,700	1,152,800	7,142,558	64.9
Austria-Hungary	7,800,000	3,620,000	1,200,000	2,200,000	7,020,000	90.0
Turkey	2,850,000	400,000	325,000	250,000	975,000	34.2
Bulgaria	1,200,000	152,390	87,500	27,029	266,919	22.2
TOTAL	22,850,000	8,388,448	3,386,200	3,629,829	15,404,477	67.4
GRAND TOTAL!	65,038,810	21,189,154	8,528,831	7,750,919	37,466,904	57.5

As reported by US War Department in February 1924. US casualties as amended by Statistical Services Center, Office of the Secretary of Defense, Nov. 7, 1957

(cited from https://www.britannica.com/event/World-War-I/Killed-wounded-and-missing)

The Unsung Heroes of World War I

THE WAR IN THE AIR
- World War I began 15 years after the Wright brothers made their first flight
- More than 65,000 aircraft were produced by both sides
- Germany built 123 Zepplin airships which carried out more than 100 bombing raids on Great Britain
- It took 5 downed aircraft or "kills" to be a "Flying Ace."
- America's top ace, Eddie Rickenbacker had 26 kills
- Germany's Manfred Von Richthofen (The Red Baron) had 80
- 500,000 carrier pigeons were used to carry messages along the front

NEW WEAPONS
- Tanks made their first appearance on the battlefield
- The self-powered machine gun was used in war for the first time; it had a range of 1,000 yards and fired 600 rounds a minute
- Heavy artillery included the French 75mm gun and the German 420mm howitzer "Big Bertha"; artillery caused 70% of all casualties

THE TRENCHES
- More than 2,500 miles of trenches were built on the 466 mile Western Front
- There was 1 soldier for every 4 inches of the trench
- The British army treated 20,000 cases of trench foot on 1914 alone

CHEMICAL WARFARE
- World War I was the first war to use poisonous chemicals
- Poisonous gas resulted in 500,000 casualties

BATTLES
- 1.2 million men lost their lives in The Battle of the Somme
- There were more than 250 ships involved in the Battle of Jutland
- Germany lost only 178 of the 400 U-boats they built but managed to sink 5,554 ships; the most famous being the Lusitania

LOSS OF LIFE and WOUNDED
- 6.6 million civilians died–2 million in Russia alone
- 65 million men fought in World War I from 40 countries and dozens of colonies - 8 million soldiers died–that's 6,000 for every day of the war
- 22.2 million were wounded

INTERESTING FACTS

- 12 million letters were delivered to the front every week
- An explosion on the battlefield in France was heard in London
- WWI sparked the invention of plastic surgery
- Blood banks were developed in WWI
- The US was only in combat 7 months
- Woodrow Wilson's campaign slogan for his second term was "He kept us out of the war. "About a month after he took office, the United States declared war on Germany on April 6th, 1917
- Four empires collapsed after WWI: Ottoman, Austro-Hungarian, German, and Russian
- German trenches were in stark contrast to British trenches. German trenches were built to last and included bunk beds, furniture, cupboards, water tanks with faucets, electric lights, and doorbells
- During WWI, the Spanish flu caused about 1/3 of total military deaths
- In early 1917, British cryptographers deciphered a telegram from German Foreign Secretary Arthur Zimmermann to Germany's minister in Mexico. The telegraph encouraged Mexico to invade U.S. territory. The British kept it a secret from the U.S. for more than a month. They wanted to show it to the U.S. at the right time to help draw the U. S into the war on their side
- King George V (Great Britain), Kaiser Wilhelm II (Germany) and Tsar Nicholas II (Russia) were cousins and grandchildren of Queen Victoria.
- A young bear cub called Winnie was one of the most popular attractions at London Zoo. He was a mascot belonging to the Canadian Army, who had been left there for safekeeping. The bear was seen by author A. A. Milnes son Christopher Robin, who renamed his own toy bear Winnie–and it became the inspiration for his dad's book, Winnie The Pooh
- The Choctaw code talkers *(above left)* were a group of Choctaw Indians from Oklahoma who pioneered the use of Native American languages as a military code. Their exploits took place during the waning days of World War I.

(Cited from https://historykids.net/history/world-war-1-facts-and-information/ and https://www.factretriever.com/world-war-i-facts and http://primaryfacts.com/1645/world-war-1-facts-and-information/ and https://www.natgeokids.com/uk/discover/history/general-history/first-world-war/#!/register and: https://www.history.com/topics/world-war-i/world-war-i-history/infographics/world-war-i-by-the-numbers#)

The Unsung Heroes of World War I

EIGHT MILLION HORSES
AND COUNTLESS MULES AND DONKEYS
DIED IN THE FIRST WORLD WAR.

Sources

Internet Sources

http://bam.files.bbci.co.uk/bam/live/content/zqn9xnb/transcript
http://birchhillmemorial.wixsite.com/ww1-ride/4-horses-returned
http://blogs.ancestry.com/uk/2011/11/07/guest-blogger-the-horrors-suffered-by-horses-during-wwi/
http://blogs.slq.qld.gov.au/ww1/2014/04/29/horses-at-the-end-of-wwi/
http://dawlishchronicles.blogspot.com/2015/01/ww1-german-view-last-years-of-cavalry.html
http://enacademic.com/dic.nsf/enwiki/4054154
http://investigatinganzacs.blogspot.com/2014/04/australian-horses-walers-of-wwi.html
http://madefrom.com/history/world-war-one/causes-world-war-one/
http://madefrom.com/history/world-war-one/role-horses-world-war-one/
http://nerdalicious.com.au/history/war-horses-britains-equine-army-of-the-first-world-war/
http://primaryfacts.com/1645/world-war-1-facts-and-information/
http://spartacus-educational.com/FWWhorses.htm
http://theanzaccall.com.au/stories/damascus.html
http://veterinarycorps.amedd.army.mil/history/ww1/ww1.htm
http://ww2today.com/the-french-sign-the-armistice
http://www.abc.net.au/local/stories/2014/08/15/4067966.htm
http://www.bahs.org.uk/AGHR/ARTICLES/58_1_5_White.pdf
http://www.bbc.co.uk/guides/zp6bjxs
http://www.bbc.co.uk/schools/0/ww1/25403869
http://www.bbc.com/news/magazine-11710660
http://www.bbc.com/news/uk-england-manchester-16970838
http://www.cbc.ca/news/canada/morning-glory-canada-s-own-wwi-war-horse-1.1259736
http://www.dailymail.co.uk/news/article-2045816/Unshakeable-courage-real-War-Horses-The-million-forgotten-animals-killed-frontline.html
http://www.dailymail.co.uk/news/article-2080777/Churchills-mission-rescue-war-horses-officials-bring-tens-thousands-home.html
http://www.dailypress.com/features/history/dp-nws-world-war-i-war-horses-1-20141129-story.html
http://www.diggerhistory.info/pages-conflicts-periods/ww1/lt-horse/first-aust-horse.htm
http://www.firstworldwar.com/features/forgottenarmy.htm
http://www.globetrotting.com.au/bill-the-bastard/
http://www.greatwar.co.uk/battles/
https://www.history.com/news/the-first-battle-of-the-marne-100-years-ago
http://www.history.com/topics/world-war-i/battle-of-the-somme
http://www.history.com/topics/world-war-i/manfred-baron-von-richthofen
http://www.historylearningsite.co.uk/world-war-one/lawrence-of-arabia/
http://www.historylearningsite.co.uk/world-war-one/the-western-front-in-world-war-one/cavalry-and-world-war-one/
http://www.homefrontfriends.org.uk/wwihorse/wwihorsedespatch/page11.html
http://www.horsetrust.org.uk/history/yesterday/war-horses/
http://www.kumc.edu/wwi/essays-on-first-world-war-medicine/index-of-essays/veterinary-medicine/horses.html
http://www.lighthorse.org.au/resources/military-stories-ww1/a-story-about-a-horse-called-jezebel-in-my-grandfather2019s-squadron-in-ww1
http://www.militarian.com/threads/wwi-sinai-and-palestine-campaign.7252/
http://www.nationalarchives.gov.uk/education/resources/letters-first-world-war-1915/horses-10000-week-come/NO
http://www.readex.com/blog/real-war-horses-america
http://www.researchingww1.co.uk/british-cavalry-regiment-1914
http://www.telegraph.co.uk/history/world-war-one/11069681/Heroic-First-World-War-horse-Warrior-receives-animal-Victoria-Cross.html
http://www.telegraph.co.uk/history/world-war-one/11202449/The-real-life-War-Horse-Cupid-the-bay-mare-from-Essex.html
http://www.todayifoundout.com/index.php/2014/03/horses-world-war/
http://www.turkeyswar.com/army/army_organization.html
http://www.warriorwarhorse.com/jack-seely.asp
http://www.wearethemighty.com/articles/these-crusader-knights-answered-the-call-to-fight-world-war-i
http://www.worcesternews.co.uk/news/9500861.Dad___s_war_horse_was_the_star_____100_years_ago/
http://www.yprespeacemonument.com/horses-and-ww1/
https://anzacday.org.au/the-light-horse
https://collection.nam.ac.uk/detail.php?acc=1994-06-217-8
https://dspace.lib.cranfield.ac.uk/bitstream/handle/1826/3032/D%20Kenyon%20Thesis%20corrected.pdf;jsessionid=F88B73FF498FA684B3294E3A3D531D11?sequence=1
https://en.wikipedia.org/wiki/German_cavalry_in_World_War_I
https://en.wikipedia.org/wiki/Horses_in_World_War_I
https://encyclopedia.1914-1918-online.net/article/eastern_front
https://forward.com/news/205157/my-opas-story-of-world-war-ones-other-fight/
https://heritagecalling.com/2017/02/28/8-memorials-to-animals-in-the-first-world-war/
https://historykids.net/history/world-war-1-facts-and-information/
https://ipfs.io/ipfs/QmXoypizjW3WknFiJnKLwHCnL72vedxjQkDDP1mXWo6uco/wiki/Italian_Campaign_(World_War_I).html
https://knowledgebank.org.nz/1703/1755/40403

The Unsung Heroes of World War I

https://mashable.com/2015/07/29/elephant-farming-war/#49M0k0_rMmq2
https://miepvonsydow.wordpress.com/2016/04/07/french-cuirassiers-only-a-year-before-wwi-would-begin-looking-much-the-same-as-they-did-under-napoleon-ca-1913/
https://nzhistory.govt.nz/media/video/bess-horse-great-war-story
https://nzhistory.govt.nz/war/nz-first-world-war-horses/egypt-gallipoli
https://nzhistory.govt.nz/war/nz-first-world-war-horses/end-of-the-war
https://parkcityhistory.org/wp-content/uploads/2012/04/Teacher-Background-Information.pdf
https://prezi.com/jyb2fiunj_zi/what-made-ww1-different-from-all-other-wars/?webgl=0
https://rarehistoricalphotos.com/italian-cavalry-school-1906/
https://schoolworkhelper.net/where-was-world-one-fought-theatres-of-wwi/
https://thetyee.ca/News/2013/03/06/Horse-Dung-Big-Shift/
https://weaponsandwarfare.com/2015/11/20/austro-hungarian-cavalry-wwi/
https://wikivisually.com/wiki/Horses_in_World_War_I
https://www.armymuseum.co.nz/museum-news/world-war-one-shared-histories/
https://www.awm.gov.au/articles/blog/the-charge-of-the-4th-light-horse-brigade-at-beersheba
https://www.awm.gov.au/articles/encyclopedia/horses
https://www.awm.gov.au/articles/encyclopedia/horses/sandy
https://www.bl.uk/world-war-one/articles/the-war-effort-at-home)
https://www.britishbattles.com/the-battle-of-the-marne/
https://www.chroniclelive.co.uk/news/north-east-news/memoirs-world-war-one-soldier-10067710
https://www.cliffsnotes.com/cliffsnotes/subjects/history/how-did-world-war-i-start-and-end
https://www.dailypost.co.uk/news/local-news/horse-penmaenmawr-who-served-first-2665570
https://www.express.co.uk/life-style/life/789444/War-Horse-Memorial-Ascot-honour-First-World-War-heroes
https://www.express.co.uk/news/history/615101/World-War-One-horses-killed-Remembrance-Day-November-11
https://www.express.co.uk/news/world-war-1/553386/Real-life-War-Horse-story-uncovered-in-animal-s-First-World-War-heroics
https://www.factretriever.com/world-war-i-facts

https://www.history.com/topics/world-war-i/world-war-i-history/infographics/world-war-i-by-the-numbers#
https://www.horsetalk.co.nz/2018/02/22/war-horses-ww1-battle-verdun-honoured/
https://www.nam.ac.uk/explore/british-army-horses-during-first-world-war
https://www.natgeokids.com/uk/discover/history/general-history/first-world-war/#!/register
https://www.nps.gov/articles/harry-s-truman-and-the-influences-of-his-service-in-world-war-i.htm
https://www.quora.com/Why-were-other-countries-involved-in-WW1
https://www.rtbf.be/ww1/topics/detail_the-horse-an-essential-participant-of-the-great-war?id=8358614
https://www.scribd.com/document/361355407/Field-Service-Pocket-Book-1914
https://www.thebrooke.org/about-brooke/history-brooke/dorothy-brookes-letter-morning-post
https://www.thedonkeysanctuary.org.uk/blog/8284
https://www.warhistoryonline.com/war-articles/researcher-debunks-russian-rumor-world-war-cossacks.html
https://www.warhistoryonline.com/world-war-i/10-facts-battle-mons-first-battle-british-german-armies-wwi.html
https://www.warhistoryonline.com/world-war-i/ed-ok-10-facts-end-wwi-armistice-november-11th.html
https://www.warmuseum.ca/firstworldwar/history/battles-and-fighting/tactics-and-logistics-on-land/supplying-war/

Books, Periodicals, Research Papers and Movies

Churchill, Winston. *The Unknown War: The Eastern Front.* C. Scribner's Sons. 1931
Galtrey, Captain Sidney. *The Horse and the War.* Country Life. 1918
Kenyon, David. British Cavalry on the Western Front 1916-1918. Thesis. Cranfiled University
Levin, Jonathan V. *Where Have All the Horses Gone?* McFarland and Company. 2017
"War Horses of WWI." Director: George Pagliero. Testimony Films. 2012. DVD
"The Walers: Australia's Great War Horse." Producer Marian Bartsch. Mago Films. 2014. DVD

Very special thanks to

Jonathan R. Casey - Director, Archives and Edward Jones Research Center National World War I Museum and Memorial–Kansas City, MO
Marian Bartsch – Producer "The Walers" Australia's Great War Horse"
Derek Donovan - Community Engagement Editor The Kansas City Star
Angelina Zaytsev The Hathi Trust
Research Facilities Imperial War Museum–London, England
and to my own horses who provided me the opportunity to love them more than usual when I needed to in the process of writing this book - MCF

The Equine Heritage Institute, Inc. is a 501 (c) 3 not-for-profit corporation. Founded in 1995, the mission of the Equine Heritage Institute is to educate, celebrate and preserve the history of the horse and its role in shaping world civilizations and changing lives.

Additional information about the Equine Heritage Institute is available via:
Website: info@equineheritageinstitute.org
Email: gloria@gloriaaustin.com

To donate to this valued not-for- profit private operating foundation:
http://www.ehi-donations.com

Our Mission Statement: To educate, celebrate, and preserve the history of the horse and its role in shaping world civilizations and changing lives.

www.equineheritageinstitute.org
www.equineheritagemuseum.com

EQUINE HERITAGE INSTITUTE

Please contact:
Gloria Austin
3024 Marion County Road
Weirsdale, FL 32195
Phone: 352-753-2826
Fax: 352-753-6186
www.gloriaaustin.com

www.ingramcontent.com/pod-product-compliance
Lightning Source LLC
LaVergne TN
LVHW071732060526
838200LV00031B/477